Our Birth On Earth

Chris Stubbs

TRAFFORD
Publishing

Note for Librarians: A cataloguing record for this book is available from Library
and Archives Canada at www.collectionscanada.ca/amicus/index-e.html

Printed in Victoria, BC, Canada.

ISBN: 978-1-4251-8588-6 (sc)
ISBN: 978-1-4251-9054-5 (e)

*Our mission is to efficiently provide the world's finest, most comprehensive book publishing
service, enabling every author to experience success. To find out how to publish your book, your
way, and have it available worldwide, visit us online at www.trafford.com*

Trafford rev. 8/8/2009

 www.trafford.com

North America & international
toll-free: 1 888 232 4444 (USA & Canada)
phone: 250 383 6864 ♦ fax: 812 355 4082

To my wife Angela
And to my daughters Helena and Hilary

FOREWORD

It can't be often that a Briton sings the praises of four French people but that is just what happens in this book. Firstly, the geocentrist, Jean Baptiste Morin, devised the House System recommended here and used throughout the book. Subsequently P. Teilhard de Chardin logically extended Darwin's theory of evolution by proposing that the drive to increased consciousness constitutes evolution's basic motivation rather than natural selection by the survival of the fittest Finally, and monumentally, Michel and Francoise Gauquelin tried to establish statistically, one way or another, the truth about natal astrology. Of course there have been many other contributors to research in astrology (Teilhard was not one of them) from several countries, to whom the references inside bear testimony, but to me the contributions made by these four French people stand out.

The purpose of the book is to kindle a spark of inspiration within a few, would-be, dedicated contributors in order to persuade them to take advantage of the rich vein of astrological opportunity laid out in the book. If it succeeds at all, then I shall be well satisfied. I can imagine, and hope, that Teilhard (despite some misgivings!) would also have been pleased.

"Know then thyself, presume not God to scan;
The proper study of Mankind is Man."

An Essay on Man: Epistle II.
Alexander Pope (1688 – 1744).

Wirral, Merseyside,
U.K.
2008

CONTENTS

CHAPTER 1

Reflections of the Moon

"In a bowl to sea went wise men three,
On a brilliant night in June;
They carried a net, and their hearts were set
On fishing up the Moon."

The Wise Men of Gotham, Paper Money Heroes.
T. L. Peacock (1785 – 1866).

Pregnant animals use the Moon to determine when to give birth. For human mothers-to-be birth occurs when the position of the Moon in the sky coincides with the Moon's reflected image of that which occurred at the Moment of Fertilisation. There are two possible plane mirrors for reflection at right angles (orthogonal) to each other. The first lies flat along the line joining the East horizon with the West, whereas the second lies vertically along the line joining the middle of the sky above the earth to the middle of the sky below it*. This conclusion comes from a theoretical treatment of the birth process (The Pre-Natal Epoch[1], see Ch. 6), but it is very tempting to suggest (dare it be said?) that it mirrors reality.

There is a related occurrence that suggests that the foregoing is possible. Life chemical compounds on the Earth show optical activity (homochirality)[2] i.e. their solutions rotate the plane of polarised light passing through them almost always to the left and only rarely to the right.

<u>Diagram 1</u>: The Moon's Position at Birth and its either/or Reflected Image Position at Epoch.

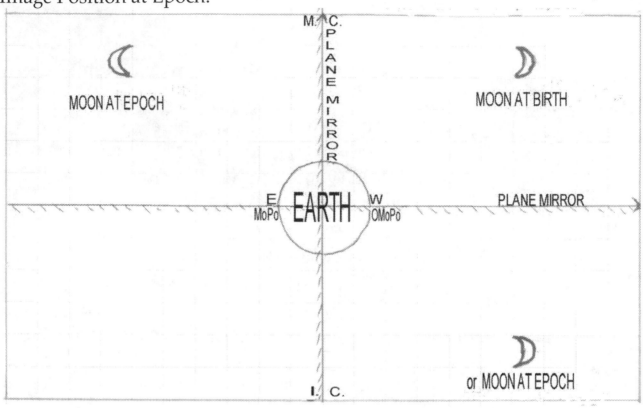

For a basic, but not from life, example of optical activity, look at:

<u>Diagram 2</u>: The Oppositely Light-Rotating Isomers of FluoroChloroBromoMethane

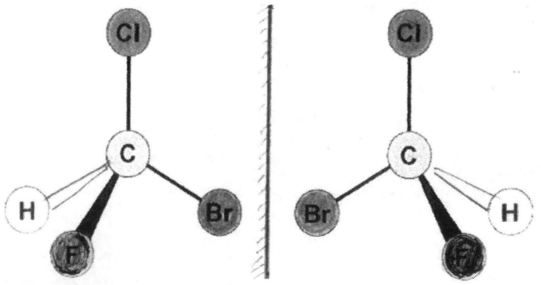

Note that these two compounds are not super imposable (just like our own left and right hands when facing each other) and so are different This means that when a plane of (polarised) light passes through each of these compounds it becomes turned to the left, or equivalently to the right.

The biological processes producing these compounds probably depended originally on the Moon circling the Earth in the direction that it does, coupled with the Earth spinning round once each day in the direction that it does (see Ch. 3). Thus this combination of Moon's motional direction and Earth's spin direction produces left, light-rotating, life compounds. Conversely if the Moon's direction round the earth had been reversed and the direction of the Earth's spin had remained unchanged, or vice-versa, then the biological process would have produced equal amounts of left and right, light-rotating compounds (or perhaps nothing at all?). If both directions had been reversed then we should have obtained life chemical compounds that rotated light to the right.

Interestingly, left-handed DNA helices (life's basic chemical compound), as distinct from right handed ones that contain more strain, and so tend not to exist, rotate the plane of polarised light to the left. This suggests that only the existing directions of the Moon's motion coupled with that of the Earth's spin (rather than their opposites) can lead to left, light-rotating compounds and so sustain life on Earth. Assuming that all of the foregoing is true then it doesn't bode well for there ever having been life on Mars, which lacks any substantial satellite!

We can get a good idea of what we have just been saying by looking closely at our own faces in a plane mirror. The image we see is not how others see us but is a laterally inverted one in that the left side of our face becomes the right side of the image's face and vice-versa. In short we have asymmetrical faces; the left side is not the identical reflection of the right. Perhaps part of the charm of our faces lies in their asymmetry! If we make a second image of the first image in another plane mirror, then we laterally invert the first image and so we now see ourselves as others see us. Take a digital photograph of your face's image in a mirror, and then take another photo of your face directly and at the same angle. Transfer these photos to your computer and duplicate (copy and paste) them. Keep one photo of each as two versions of your asymmetrical face and, after resizing, divide each of the other two, into two (after another copy and paste) vertically and equally down the middle. Combine the left side of one with the right side of the other and also the right side of the one with a left side of the other. In this way we create two different, symmetrical faces of ourselves. In all we have four different faces, each one of which could conceivably have been ours, but only one asymmetrical one is ever the true one (see pictures).

<u>Adam's four faces:</u>

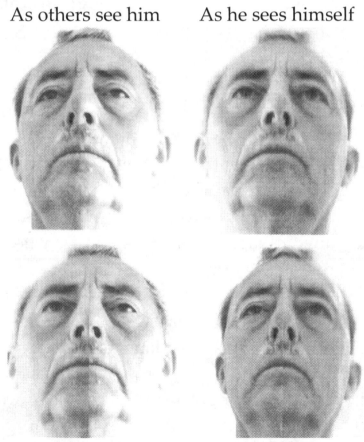

As others see him   As he sees himself

Two extra symmetrical faces

In a related way, the biological processes leading to light-rotating, life compounds essentially only ever produce the one that turns plane (polarised) light to the left.

Michel and Francoise Gauquelin (The Gauquelins) showed (among several other things) that, at the times of birth of eminent writers, the Moon was rising over the Eastern horizon significantly more often than expected by chance. A similar but less pronounced effect was noticed for the Moon passing over the middle of the sky above, at the times of birth[3]. Characteristics associated with the Moon, and derived empirically by Hankar[3], include: energetic but not athletic, strong willed, tenacious, practical, egoistic, initiating but not duly ambitious, moderately intellectual, realistic imagination, little erotic and/or romantic inclination. A more generous set of characteristics includes: charming, gracious, sympathetic, kind and showing sensibility. We should expect these characteristics, by and large, to feature strongly in the personalities of top writers.

- -

References and footnotes:

* Actually, the orthogonal mirrors lie in the plane of the Zodiac; their reflective faces are both perpendicular to this plane.

1) The Pre-Natal Epoch, E. H. Bailey, Samuel Weiser, New York, USA, 1974.

2) Wikipaedia, The Internet, Chirality and Homochirality.

3) Recent Advances in Natal Astrology, G. Dean and F. Mather, The Astrological Association, Bromley, Kent, UK, 1977.

- -

CHAPTER 2
The Science of Astrology.

"Astrology supplies Biology with the Precision of Physics"

In Chapter 1 we started talking about science matters and mentioned three research Astrologers. Before proceeding further with science affairs, and where Astrology fits in, let us consider science and the four main sciences briefly in order to see what science is about and what one or two of the basic differences between the four main ones are.

Since we were youngsters we have often come across the word "science". But just what does this word mean? A common definition says that "A science is an organised body of knowledge (e.g. a Physics textbook), that is capable of quantitative treatment." The keyword here is "quantitative" although the word "organised" is also important. Now let us look at the four main sciences:

1) <u>Mathematics</u> is a pure science of mental constructs only into which actual matter doesn't intrude at all. As a result we can specify mathematical constants as precisely as we should like. For example the constant π, by which we multiply the diameter of a circle in order to obtain that circle's circumference, has a value of 3.1415926536. . . <u>Logic</u> is another example of a pure science.

2) <u>Physics</u> is a natural science and involves the properties of matter. Because there is a limit to the extent to which we can define/measure matter we cannot specify physical constants to quite the same degree of precision as we can mathematical ones. <u>Astronomy</u> is a branch of Physics.

3) <u>Chemistry</u> is another natural science, but because the matter of chemistry is often being changed from one chemical form into another, we can't specify chemical constants to quite the same degree of precision as we can physical ones.

It has always seemed incredible to me that common salt (sodium chloride), that I can shake onto the palm of my hand, here on Earth, is exactly the same as that I could handle in just the same way were I to live on a planet of a star in the constellation of the Great Bear, i.e. the same matter is identical, under the same conditions, throughout the entire Universe.

4) <u>Biology</u> is an example of a life science, presently earthbound, but because the matter of biology is extremely complicated when compared with the inanimate matter of Physics and Chemistry, and also because water is always associated with it in significant amounts, we find it relatively difficult to specify biological constants precisely. Nevertheless almost everyone agrees that biology, and its branches (e.g. zoology, botany) are indeed true sciences.

Suppose I drew a short, straight line on a piece of A4 paper and photocopied it directly several times. I could send a copy to different people in different parts of the world and ask them to measure carefully the length of the line in centimetres and then send the result back to me. If I then measured the length of the line myself and found it to be 3.7 cm. long, I should be confident that the responses that I received would all be very close to 3.7 cm. Measurements such as these that are independent of whoever it is that makes the measurement, or of wherever they are, or of whenever they do it, are said to be "objective". On the other hand those measurements (or assessments) that can vary depending on who is making the measurement/assessment, are called "subjective". Hence we can extend our initial definition of a science by the word "objective", so that it now reads, "A science is an organised body of knowledge that is capable of objective, quantitative treatment."

Consider now an example of a subject that is not a science, such as <u>Art</u>. We say that beauty lies in the eye of the beholder. This means that assessments of beauty are "subjective", and so rather difficult to quantify.   Thus   we   could   define   Art   as   a   body   of knowledge/experience that is capable only of subjective, or qualitative, treatment. But we need to be just a little bit careful here. Our subjective assessors of beauty are all human beings. Although they are all very much alike, they are, at the same time, undeniably different from one another. It is likely that people differ only in degree, rather than by using different mechanisms or processes. Hence the various factors that contribute to each one's assessment of beauty could themselves be quantifiable, provided that we knew enough about them. Conceivably then, a subject such as Art could eventually become a science.

Now let us take a look at <u>Psychology</u>. Here we have a subject that deals with living things (e.g. ourselves) and so our first impression is that it can only be, at best, a relatively imprecise science. When we then consider that it deals with our personality characteristics, which are clearly subjective (we don't say that Bill Bloggs is 0.75 aggressive!), we need to question whether we can classify Psychology as a science at all. People are divided on this issue. Those in favour of it being a science classify it, not as a branch of Biology, but separately, as one of the behavioural sciences. Over the years people in general have come to accept that Psychology contains truths that are useful to society. Certainly it uses scientific methods, such as statistics, to derive many of them.

So, in the light of the foregoing, how do we begin to classify Astrology? Electional and Horary Astrology, much of Mundane and Predictive Astrology, that involve events, beginnings, questions, politics, countries, world affairs and irregular time intervals, appear to lie beyond the scope of science. On the other hand we can with ease, assign to several different sciences, cyclical studies that relate the periodicities of the Sun, Moon and planets to people, animals, economics and the weather.

In between these two extremes stands the case for Natal Astrology. This seems very similar to that for Psychology. Here we are telling people about their own personality traits, relationships, careers and

14

health. Thus we could well classify Natal Astrology as a behavioural science. However, the setting up of an astrological chart centred on the Earth is very much a branch of Physics, and this part of Astrology could well belong to the science of Geocentric Astronomy.

Regardless of whether or not we are willing to classify Natal Astrology as a science, science and its methods currently play an important role in trying to establish the truth of Natal Astrology. Budding astrologers, having interpreted the horoscopes for their relatives and friends, have little difficulty in accepting its truth. But many people can't imagine how Astrology can possibly work. They require answers to questions such as "How on Earth can the planet Venus impart different harmonising characteristics to different people's personalities depending simply on the time when, and the place where, they were born?"

To begin to answer this deceptively simple yet extremely difficult question let us take a look at the:-

Physical Evidence for the Truth of Natal Astrology.

Tomaschek[4] stated that there are, beyond doubt, correlations between the positions and aspects (i.e. the specific angles 0, 30, 45, 60, 90, 120, 135, 150 and 180 degrees subtended at the Earth by two different planets) of the planets, and angles (i.e. the Eastern horizon and the middle of the sky above) with terrestrial events. He listed four possibilities for explaining such correlations:-

1) Celestial bodies actually OPERATE upon terrestrial events.
2) Celestial bodies PRECIPITATE such events.
3) Celestial bodies SYNCHRONISE with such events.
4) Celestial bodies SYMBOLISE organic cosmic forces.

Possibility 1) is unlikely due to the small energies involved. Roberts[5] has strongly and adversely criticised possibility 3) and possibility 4) could prove to be a higher order extension of possibility 2). So for the moment we shall concentrate our thoughts and efforts on possibility 2). Here, like possibility 1), energy is also involved, but it is a minimum value required for initiation. This then is hugely magnified by resonance and finally given the necessary selectivity, or variation, depending upon the frequencies involved.

Using statistical methods the Gauquelins rigorously showed that certain planets, in prominent positions with respect to the Earth (i.e. coming up over the Eastern horizon and just passing the middle of the sky above), occurred there significantly more often than chance, for specific elite groups[3]. Ertel[6] has confirmed this. Similarly Dieschbourg[3] demonstrated less rigorously that planetary aspects could also be related to elite, professional groups. Addey[7] succeeded in rationalising the Gauquelins' results and could, presumably, have rationalised Dieschbourg's also, using harmonic analysis. This possesses the potential to express the complex requirements of the genetic code adequately in astrological terms, whereas traditional analysis by Zodiacal signs, by Houses of the horoscope (that describe our earthly experience) and by conventional aspects, although extremely useful and startlingly accurate at times, lacks the scope to do so. Later Addey made a start on building sets of individual interpretations by detailed examination of the 4th harmonic of several planets. On the other hand there is little statistically significant evidence for the existence of Zodiacal signs, or for the Houses of the horoscope.

In an attempt to answer the question, "What is the most correct rising point for planets?" Stubbs* criticised the traditional point (the Ascendant) and in its place recommended the Morin point within the Morinus House system, for its simplicity and world-wide applicability.

Roberts[5] has pointed out that these planetary influences must travel at the speed of light. Seymour[8], while acknowledging that the influences are weak, claims that planetary gravitational pull on the Earth's upper atmosphere (as the Moon's gravity helps to generate the tides on Earth), while also extremely weak, can undergo huge (10,000 X) resonance amplification, that, in turn, can disturb the Earth's magnetic field, just as the Moon can. That the planets are capable of disturbing the Earth's upper atmosphere in this way is shown analogously by their known influence on sunspot cycles. The planets' combined gravitational pull, particularly when in hard aspect (i.e. 45, 90, 135 and 180 degrees), causes their irregular behaviour. The geomagnetic disturbances, of the appropriate frequencies, then affect our nervous systems (MIDDLE GROUND). Seymour has shown that his theory can predict the relative variations of the planetary influences, from their point of closest approach to the Earth, compared with that when they are furthest away. His theory can also explain the observed effects for geomagnetically disturbed days.

<u>Diagram 1</u>: The Physical (Astrological) Side.

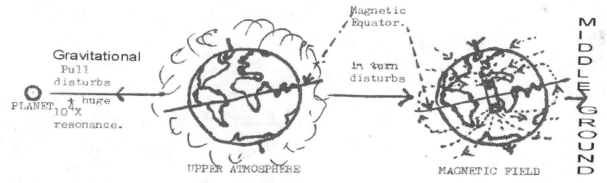

The Middle Ground

Making the assumption that we are still on the right lines, what we now have to ask is how our central nervous systems detect the magnetic disturbances directly or indirectly. Taking the direct case, F. A. Brown[9] has reviewed how animals respond to magnetism, and Seymour[8] has briefly updated this.

Taking the indirect case, Seymour[8] suggests that fluctuations in the Earth's magnetic field have electric currents associated with them that can cause nerve impulses to pass along the nerve cells that in turn are detected by our central nervous system, in a similar way to how a T.V. aerial plus tuned receiver picks up radio waves. Along these lines Roberts[5] has proposed that an intangible, flower-like organ senses the magnetic disturbances, apparently in the form of psychic waves that somehow become manifested as our personality characteristics. Alternatively, magnetic disturbances can alter the degree of rotation of plane polarised light. Possibly it is this that we detect with our eyes, presumably in an involuntary fashion.

We have now presented the physical evidence and have introduced the middle ground. Let us now consider the biological side.

- -

References and footnotes: (lower numbered references occur at the end of Chapter 1.)

* C. Stubbs, Astrology Quarterly, <u>64</u>(4), 27 (1994). See also articles published by C. Stubbs in Urania, Utrecht, The Netherlands, 1995 – 2007.

4) R. Thomaschek in a "New Study of Astrology", J. M. Addey, Urania Trust, London, 1996, Appendix 3.

5) The Message of Astrology, P. Roberts, The Aquarian Press, Wellingborough, UK, 1990.

6) The Tenacious Mars Effect, S. Ertel, Urania Trust, London, 1996.

7) Harmonics in Astrology, J. M. Addey, L. N. Fowler, Romford, UK, 1976, Ch. 20

8) Astrology: The Evidence of Science, P. Seymour, Leonard Books, Luton, UK, 1988.

9) How Animals Respond to Magnetism, F. A. Brown, Discovery, November 1963.

- -

CHAPTER 3

In the beginning . . . . . .

"The Name of the Game is Astrology,
The Name of the Game is DNA[10],
The Name of the Game is Life."

"For Life, the Earth is the Centre of the Universe."

The Earth-Moon binary system was created some 4.5 billion (thousand million) years ago probably by a large collision between the original (proto) Earth and a Mars sized body (Theia) having an orbit around the Sun similar to that of proto-Earth[2]. We think that this collision occurred on the offside (from the Sun) of proto-Earth causing the resulting proto-Earth – Theia combination to spin in the direction that it does, and ejecting the Moon to move off in the plane of the Solar system (i.e. in the plane of the ecliptic) rather than in the plane of the equator (i.e. the central plane of the Earth's spin) and in the direction around the new Earth that it now does. We explain the fact that the Moon's core is relatively small compared with that of the Earth because the offside collision made it consist more of proto-Earth's, and Theia's crusts, than of their central cores.

Without the Moon and so without its role in generating birth processes, it seems unlikely that life on Earth would ever have evolved as it has. A criterion such as this, for evolved life as it is today, from its earliest beginnings, constitutes a Prime Cause[4]. There are other Prime Causes for life's evolution on Earth, for example 1) the Sun for its heat and light; 2) the availability of water over most of the Earth's temperate surface; 3) liquid water being more dense than its

<u>Diagram 1</u>: The Production of the Moon and the Earth from the Collision of Theia with Proto-Earth.

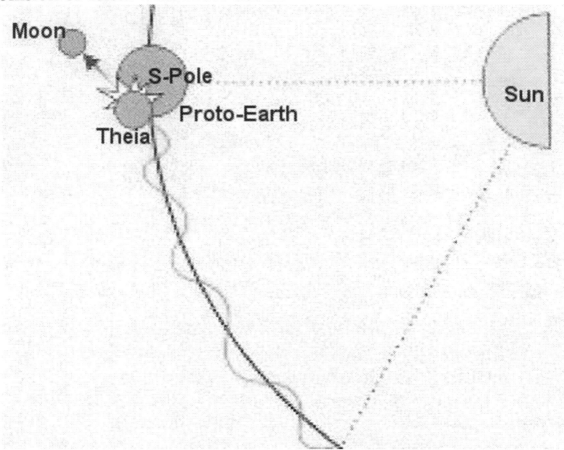

solid form, ice (this is rare!), so that it freezes from the top down enabling life to function in the cold, liquid water beneath; 4) the presence of chemical elements essential for life, such as carbon, sulphur, nitrogen and phosphorus and 5) mutations (see following).

We know that life has left traces of its existence here for almost four (3.8) billion years[5]. Thus the Earth has spent relatively little time (only 700 million years!) as a lifeless planet, following its binary creation with the Moon. The billions of years that have elapsed since the Earth first held life comprise the "deep-time" (compared with "deep-time" all of human time seems but a passing few moments) of joint biological evolution and geological change. We are very familiar with the latter from television documentaries showing grinding tectonic plates and the accompanying earthquakes, belching volcanoes, landslides and tsunamis that rush across the oceans. Additionally, we must not forget the more usual ravages of fire, storm, tide and wind, all of which contribute to massive geological change. By contrast biological development (fertilised cells -> baby, etc.), if not evolution, requires calm, controlled, stable and safe conditions in order to flourish. If it is true that development depends in part on reflections of the Moon, then presumably conditions would have needed to be particularly calm over prolonged periods of time for this dependency to develop and so be able to signify the end of the birth process.

We can visualise the Earth as a (bio)sphere covered with a thin skin of tissue called life[11]. Living things are composed of invisible, soft, building blocks called cells, every one of which carries within itself a singular chemical called deoxyribonucleic acid (DNA). DNA is a large (life) chemical compound (molecule) made from just five chemical elements, namely carbon, hydrogen, oxygen, nitrogen and phosphorus. How a life compound as complicated as DNA managed to evolve and replicate the way that it does and did, is a mystery with which our boggling minds are barely beginning to get to grips.

DNA alone unites all life in a common history because every cell of every living thing has contained a version of DNA for a billion years. Because DNA can replicate itself, living things can produce offspring and so possess a common descent from shared ancestors. The breathtaking idea that a single DNA based life-form was the ancestor of all living things spawns a sort of "Big Birth" theory.

The growth of cells from a fertilised egg into a living creature is called development and the development of life on Earth is called evolution. Both development and evolution bring about structures of amazing complexity that are time-dependent and structured hierarchically, from the interaction of genes (entirely composed of DNA) with proteins. The DNA in every cell of a person – called that person's genome – is very like an encyclopaedia in design and content. We ourselves are very big compared to a cell, and cells are very big compared with the atoms of the chemical elements from which they, and so we, are made. We are composed of one hundred trillion (million, million) cells and each cell is made up of one hundred trillion atoms. Thus the complexity of a cell in atomic terms is about as great as the complexity of a person (brain included) in cellular terms.

Interestingly 99.6% of our working DNA is just the same as that of a chimpanzee[12]. That residual 0.4% suddenly seems remarkably important!

Mutations.

Mutations are the sudden appearance of a new, variant form, or behaviour (as in personality characteristics?) that does not go away but rather breeds true in successive generations. These new forms, or patterns, are caused by changes in the sequence of bases (mainly four) in the DNA of a gene. The rigid, base-pairing rules, that enable information to be copied from one DNA double helix into two, can also fix in place any error that occurs. Even a single base-pair, if copied or repaired wrongly, just once, can completely change the meaning in all subsequent generations of that DNA.

Mutations are rare, but beneficial ones, since life is old, are the underlying cause of life's diversity. Without mutations life would have died out long ago from its failure to adapt to fluctuations in temperature, atmosphere and water level. Hence mutations constitute another Prime Cause. With slight, but continuous, mutation the descendants of some creatures have been able to survive myriad environmental disturbances to become the millions of different species that we recognise today.

Looking at this present richness of life on Earth it is difficult to believe that natural selection, that permits the survival of some, but not all, randomly occurring sequence differences in DNA, is responsible for so many different life forms. But every mutation can produce a new form, and each mutation must make sense in its own context, before it can serve as the new base line for the next mutation. In this way a series of changes will accumulate over time that will be seen with hindsight, but only with hindsight, as a remarkable parody of an intelligent plan. DNA's wasteful, but so far successful, strategy for surviving environmental stress and competition, through imperfect replication, drives Darwinian natural selection.

On the other hand some changes can be harmful. Causes of cancer, such as tar from cigarettes, particular chemicals (carcinogens) as well as sunlight and other forms of radiation, can lead to complete breaks in DNA. Breaks in genes occurring in cells that will merge to begin a developing embryo, can be fatal, or often result in serious illness from birth on.

But for us, the rarity of general mutations means that, for all practical purposes, the DNA of the first, fertilised egg-cell, and that of all the cells of the resulting new-born baby, will be identical.

Biological Clocks[13].

Studying Astrology gives us a feeling for time. All living things exist in both time and space. They function in time by using a variety of biological clocks, i.e. physiological processes that measure time using environmental cues such as the tides and light/dark cycles. These in turn derive from the Earth's moving relationship with the Moon, Sun and other members of the Solar System. Regarding Natal Astrology our attention is focussed on developmental clocks. Primarily these depend on our central nervous system (the brain, its stem and the spinal chord) and then on our hormone supplying endocrine glands, i.e. the pituitary and hypothalamus glands, that trigger the testes, ovaries and then the placenta to regulate pregnancy.

Just what are the environmental cues here on Earth that serve to regulate the developmental clocks in mammals? Possibly the answer lies with the frequencies of the magnetic disturbances produced by the resonance enhanced tidal effects of the planets on the Earth's upper atmosphere (see Chapter 2). But before proceeding further let us now look at:

The Initial Stages of Our Birth Process.

Biologically, we can say that for every individual there are five steps that lead up to, and include, conception (see the Scheme)

Diagram 3: Simplified Scheme for the Initial Steps of Our Birth Process.

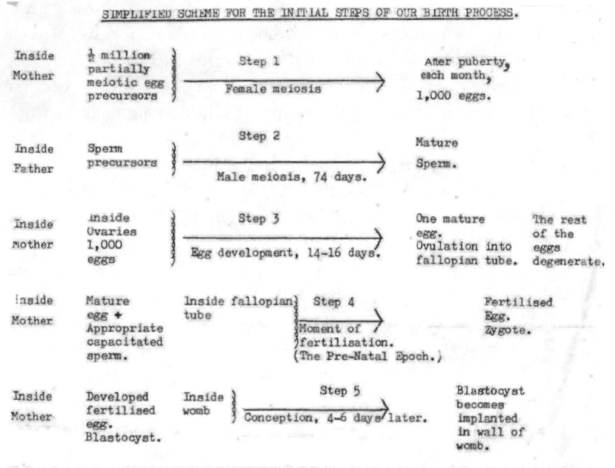

SIMPLIFIED SCHEME FOR THE INITIAL STEPS OF OUR BIRTH PROCESS.

Within these five steps (say between steps 3 and 4) there is the mutual selection of the father and of the mother. The conscious result of the mother's DNA attracts the conscious result of the father's DNA, and <u>vice-versa</u>, in the process called love, or the selection of an appropriate mate. Ideally love occurs when each can say to the other

"It is not so much the way you look, or what you are, although these are important,

But rather the way I feel, or how you make me feel, when I am with you."

24

Love is important for the rearing of children because not only will you love those parts of the child that derive from yourself, but also those parts that derive from your loved one. The denial of this love selection process is perhaps one of the main reasons why most women find rape so abhorrent.

In steps 1 and 2, female and male meiosis leads to a multitude of eggs and sperm respectively, each different from one another, and yet each deriving from mother or father through multi-consecutive division by two. During step 3 roughly 1,000 eggs, out of an original half million or so, start development each month, from which only one, in general, becomes released from one ovary or the other. Similarly in step 4, only one, out of some 300 million sperm, takes an active part at the moment of fertilisation. Even then there is a final screening in step 5 (conception), since not all fertilised eggs become attached to the wall of the womb.

Within these five steps the only obvious times necessitating planetary selection occur within the female meiosis (step 1) and within the egg development (step 3) stages, a fortnight or so before the epochal act of fertilisation (step 4). Hence the egg with the best gene fit for the characteristics of its time (given by the prevailing planetary configurations at the moment of birth) reaches maturity. Apparently nature, by necessity, is exceedingly wasteful of its genetic material in its efforts to produce one timely, new-born baby. Its approach for providing a timely birth is to swamp the situation with huge numbers. Thus the mature egg for the month has arrived by means of a two stage selection process from the mother's original egg-bank of half-a million. The father then supplies a vast number of sperm from which the pre-selected egg selects the first one with a correct gene-fit at the crucial time (step 4, the moment of fertilisation, or the Pre-Natal Epoch). At the moment of fertilisation not only is there a combination of the mother's genes with those of the father, but the sex of the baby-to-be is also determined. It is the father's invading sperm that carries the determining sex-chromosome. By the time of conception (step 5), when the wall of the womb either accepts or rejects a developing zygote (blastocyst), there appears to be no basic need for planetary influences; by this stage the genetic make-up of the potential individual is already well and truly complete.

Considerably predating the moment of fertilisation (step 4), and indeed the whole scheme, is the time when our own mother's eggs became established in her ovaries when she herself was but a ten week old foetus in our maternal grandmother's womb. Even at this foetal stage eggs begin meiosis that is halted, before our mother's birth, at what is called the dictyotene (advanced diplotene/prophase) stage. Since any astrological significance here must account for all future offspring, some maybe decades hence, it is tempting to suggest that this can only involve the outer or "generation" planets. These foetal eggs then await their mother's adulthood, and further timely stimulation (involving "personal" planets, e.g. the Moon?), in order to develop into mature eggs that are then ready to initiate the next generation (us) at our moment of fertilisation (step 4).

Similarly, sperm that developed in our father's testes derived originally from primordial cells that migrated to his testes at an early embryonic stage in our own paternal grandmother's womb. Unlike eggs however, sperm are continually generated in vast numbers, each one reaching maturity after about 74 days (⅕th of a year)[14]. As they have no long, independent existence the meiotic process (involving the "personal" planets?) that produces them is relatively recent.

Thus we can see the links, <u>particularly down the female lines</u>, across three generations.

Once conception (step 5) has taken place, the development of the embryo, and then of the foetus, takes place rapidly, like clockwork. By the time that we are ready to be born (and the foetus itself appears to play a part in this) we are very much adult creatures in miniature.

Diagram 2 reviews the immediate biological sequence presented in the last five sections:

Diagram 2: The Immediate Biological Side

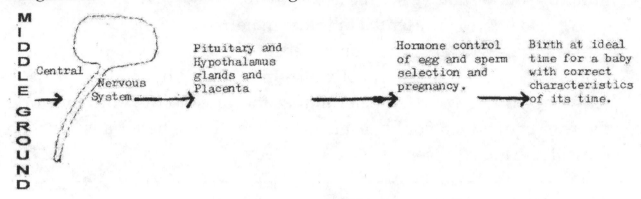

Reconsidering the middle ground within which we detect the magnetic disturbances it remains to be seen whether or not humans can detect small changes in the Earth's magnetic field directly and then make use of them for egg and sperm selection. So how does the middle ground detection process twice help to select the eggs and perhaps once the sperm, that have the characteristics right for their time? Because there is the need to select at two different stages the selection process probably has to be more developed in women than it is in men. We can account for this genetically and in general by pointing out that women have the double X chromosome whereas men only have the less complicated XY.

Diagram 4 summarises the Middle Ground:-

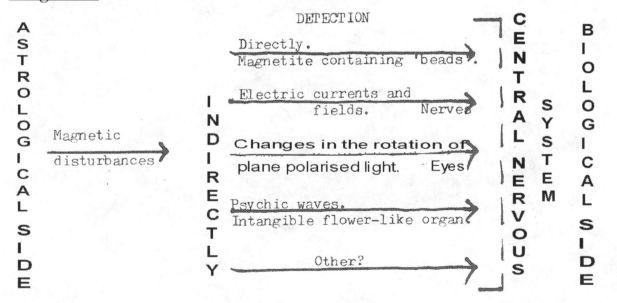

Whichever detection method applies it has become clear that it is the middle ground that we are the least sure of.

We have now presented all the material evidence required to answer the awkward question asked at the end of Chapter 2. This evidence is incomplete and not well substantiated but it does provide us with a framework for determining the direction of further work, the results of which could enable us to formulate a better answer than is available to us presently.

References and footnotes: (lower numbered references occur at the end of previous chapters)

10) Recombinant DNA, J. Watson, J. Tooze and D. Kurtz, Scientific American Books, New York, USA, 1983.

11) Abstracted and modified from "Signs of Life", R. Pollack, Penguin, London, UK, 1994

12) Shadows of Forgotten Ancestors, C. Sagan and A. Druyan, Random House, London, 1992.

13) Biological Clocks, J. Brady, Arnold, London, 1979.

14) Human Heredity, E. J. Garner, J. Wiley, New York, USA, 1983.

CHAPTER 4

The Inner Map

"His mind his kingdom, and his will his law"
W. Cowper, *Truth.*

In the previous chapter we traced a history of life mainly through DNA. Although we did mention human love as a means for selecting an appropriate mate our focus remained almost exclusively on matter. However, astrology deals mostly with our psychological make-up, i.e. our "within". In his book "The Phenomenon of Man" Teilhard de Chardin[15] proposes logically that evolution proceeds by development of mind/spirit through the increased complexity of matter, rather than through Darwinian "survival of the fittest". In animals/mammals consciousness shows itself instinctively but humans, uniquely, have gone a stage further in that they can think well constructively. Individuals collectively contribute to the noosphere, the Earth's consciousness, as the next superior layer to the biosphere. Further rises of consciousness will come as mankind manages to unify itself into ever more complicated societies/arrangements leading finally to the Omega point.

Somewhat surprisingly the role that the Moon plays is completely neglected in the whole treatise and apart from a brief mention once or twice, so are the parts played by the other members of the whole Solar System. These help us to describe our individual consciousnesses. In Appendix 5a we show Prince Harry's horoscope and present its interpretation in part. As a way of introducing the interpretation we could say:

Dear Prince Harry,

Heaven's Message

The signs of the Zodiac are of varying importance in different birth charts. Media Astrology has emphasised that Sign containing the Sun, since we can all know our own sign from our individual dates of birth[16]. However, serious astrologers know that the Sign ascending over the horizon is at least of equal importance. But this ascending Sign can only be found when the moment of birth is known, and is therefore specific for the individual for whom the chart is drawn up. In the way that the Sun is an important planet for men so the Moon is an important planet for women, although both are of importance for both sexes. The remaining planets (Mercury, Venus, Mars etc.) represent different parts of each person's consciousness, the interpretation of which varies depending on the Sign that contains them and on the aspects from other planets that each receives. All told the whole interpretation provides us with a good description of each person's character based on the moment of birth.

Please find enclosed your complete birth chart together with three or four pages of simple analysis. Note that none of the interpretation is mine; I have simply taken the various indicators and have tried to blend them into a readily understandable whole. Specifically your own birth chart shows that:

The Sign ascending at birth is Aquarius.

The Sign containing the Sun is Virgo.

The Sign containing the Moon is Taurus.

Hence you could describe yourself astrologically as an Aquarian man with the Sun in Virgo and the Moon in Taurus. Hence you are mostly a mixture of the traits associated with Aquarius, Virgo and Taurus[17].

"The ideal should never touch the real."

Schiller, *To Goethe*

"A place for everything and everything in its place."

Samuel Smiles, *Thrift*

"Let us make hay while the Sun shines."

Cervantes, *Don Quixote*

Probably it is best that you yourself should not judge the interpretation but rather let someone who knows you well judge it with you. Basically horoscopes attempt to supply the requirements for satisfying the old Greek dictum, "Man, know thyself" (presumably both individually and collectively), and so provide one of the few "inner" (within) maps available.

The general idea is for <u>you</u> to decide what suits <u>you</u> best, to build on <u>your</u> strengths, to guard against <u>your</u> weaknesses and to reinforce your own personal judgement. In this way the horoscope, along with its interpretation, tries to be useful.

- -

Overall the whole horoscope describes one single human consciousness.   Considering the several billions (thousands of millions) of people living on the Earth presently, can we possibly envisage a way of describing the collective noosphere astrologically?

Teilhard says that, *"It is the process of groping combined with the two fold mechanism of reproduction and heredity (as the hoarding and additive improvement of favourable conditions obtained, without the diminution, indeed with the increase, of the number of individuals engaged) which gives rise to the extraordinary assemblage of living stems forming the tree of life."* Astrogenealogy (see Chapters 8, 9 and 10) may help eventually to throw some light on how this process is accomplished. But let us now look at astrology's contribution to our birth on Earth.

- -

References and footnotes: (lower numbered references occur at the end of previous chapters.)
15) The Phenomenon of Man, P. Teilhard de Chardin, Rome, 1948, Harper-Torch, USA, 1965.
16) The Modern Textbook of Astrology, M. E. Hone, L. N. Fowler, Romford, UK, Rev. Ed., 1978.
17) Astrology: How and Why it Works, M. E. Jones, Routledge and Kegan Paul, London, UK, 1977.

- -

CHAPTER 5

Great Circles and the Morinus System of Houses.

"Surely we are Earthlings first, and then members of the Solar System"

Inside a planetarium, imagining that we are seated at the Earth's centre, we can look up at the domed ceiling of stars above us, and at the same time, imagine the continuation to another dome of stars beneath us. This whole sky of stars that surrounds us is called the Celestial Sphere. The plane of any circle that contains the centre of the Earth at its centre is called a Great circle. For example the Equator is a Great circle and corresponds to latitude 0^0. All other circles of latitude, parallel to the Equator, but whose plane cannot contain the earth's centre, are small circles. On the other hand, if we think about it for a moment, all circles of longitude are indeed Great circles.

Diagram 1: The Celestial Sphere (described for Polar Elevation of $53^025'N$, the Latitude for Liverpool).

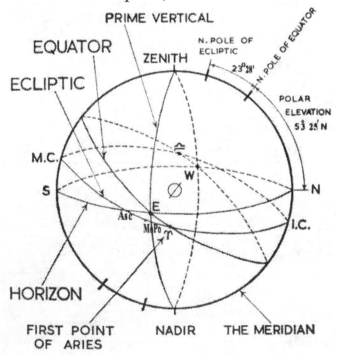

The diagram shows the major Great circles of the Celestial Sphere[18]. Probably the Meridian is the most important Great circle. It is the North – South Great circle and it passes through the North and South poles of the Equator, the North and South poles of the Ecliptic, the North and South points of the true Horizon, as well as the zenith and nadir (the poles of the true Horizon). The Sun crosses the Meridian at midday (anywhere on Earth) and this point of intersection with the Ecliptic (the Sun's apparent path and the circle of the Zodiac) is known as the Midheaven. We can speak of three other Great circles: the Prime Vertical, the Equator and the true Horizon, as secondaries to the Meridian because they all pass through the East – West points, the poles of the Meridian. The true Horizon must not be confused with the visible horizon, formed by the apparent meeting of the Earth with the sky, which is a small circle that is parallel to the true Horizon.

The Morinus House System.

Keeping our introduction of Great circles in mind we can construct our own personal map of the Zodiac, containing the heavenly bodies of the Solar System (even the Sun and Moon are considered to be "planets" of the Earth as far as Astrology is concerned – remember that the Earth is geocentric!). This is our horoscope i.e. a picture/chart of the heavens as they appeared at the time and place of our birth on Earth. Nowadays we can do this easily using a computer and the appropriate software[19]. However the Appendix 5a gives full details of how we can do this from first principles[20]. Briefly, the first step in making/casting/erecting the chart is to calculate our Star (sidereal) time of birth, unlike clock time, that is determined by the Sun. Anyone can do this by means of a very simple formula (see Appendices 5a and 6). The second step is to calculate the twelve "Houses of Heaven". Houses are different from Signs and should not be confused with them. The Signs come from dividing the apparent, circular path of the Sun around the earth once a year, by twelve. However, the Houses derive from dividing by twelve the apparent, circular path the heavens make around the Earth every day (due to the Earth's spin). Thus Signs come from the yearly cycle and the Houses from the daily cycle. By superimposing these two cycles on one another, with the Earth at the common centre, we construct a chart of the heavens for each of our birth times. Thirdly, we use an ephemeris to put the planets in the Zodiac, and finally list the various aspects (angles) they make with the Earth among themselves (see Appendix 3a). Note however, that there are in fact several different ways of making the twelve-fold division of the Zodiac into Houses. These methods all have their supporters and detractors but the method that we shall use was devised by the great French astrologer and mathematician, Morin de Villefranche, known as Morinus. His astrological logic proceeded as follows:

Our Earth itself is defined by the Equator with its axis passing through the North and South poles. At our moment of birth we can fix the position of our Houses by seeing where the Meridian and the true Horizon intersect the Equator. We trisect the quadrant so formed so that, with all four quadrants we now have twelve equidistant points around the Equator. We take these points as centres (rather than the boundaries) of the Houses, thus making them more symmetrical about the place of birth.

To specify House centres in terms of the heavens we now have to apply the sky to the Earth. We define the sky by the Ecliptic (the Zodiac), the yearly apparent path of the Sun around the Earth, with its axis passing through its own North and South poles. The points where six Great circles, each passing through the North and South poles of the Ecliptic, and through two different and opposite House centre points on the Equator, describe the House centres in terms of degrees of the Zodiac (see diagram 2). In other words we measure House centres on the Equator in degrees of celestial longitude. Essentially this is the House system proposed by Morinus. The first House centre, called the Morin point, is the East point (i.e. the point of intersection of the true Horizon with the Equator) projected onto the

Ecliptic. In a sense it is a triple E point, where the East point of the Horizon on the Equator is defined by the Ecliptic. The Morin point is at right angles (orthogonal) to the Midheaven*.

Diagram 2: The Morinus House System and its House Boundaries at Birth[+].

.... House Boundaries

– – – – – Angular House Centres

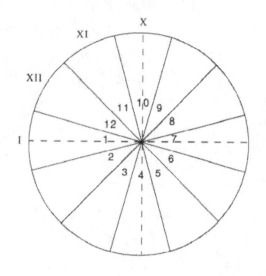

Figure 1 Distribution of House Boundaries
About the Place of Birth

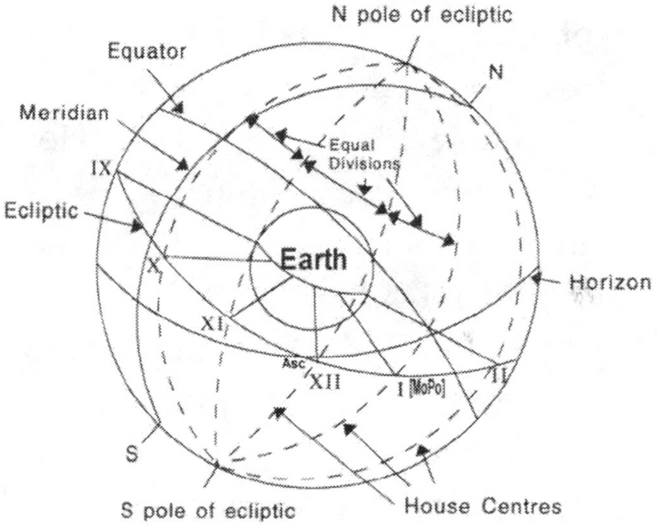

Figure 2 The Morinus House System[3]

On the other hand the traditional Ascendant is the point of intersection of the true Horizon with the Ecliptic and is the starting point for several other House systems. The question that remains is "Which of these two starting points is likely to be more correct?**" Unfortunately preliminary statistical tests (see Appendix 3b) to try to differentiate between them have proved inconclusive. However, further similar studies of births at high latitudes potentially provide a means of effecting this. As latitudes increase, the differences between the Morin Point and the Ascendant become more exaggerated.

Now we are ready to examine how astrology treats our birth on Earth.

References and Footnotes: (lower numbered references occur at the end of previous chapters.)

* For Northern hemisphere births the direction of the equator is South so that the top of the page of the horoscope has this direction and its left-hand edge represents the East. For Southern Hemisphere births the Equator's direction is North so that the top of the page now stands for this direction and its right-hand edge represents the East. Thus the horoscope for the Southern hemisphere is a mirror image of that of the Northern one about the M.C. – I.C. axis (i.e. one half rotation about this axis). The House System we use in the Northern Hemisphere in this book is that due to Morinus; its counterpart we could use for the Southern Hemisphere is called the Zariel Axial Rotation House System.

18) The Astrologer's Astronomical Handbook, J. Mayo, L. N. Fowler, Romford, UK, 1982.

19) "Solar Fire", Astrolabe Inc., Brewster, Maryland, USA,

20) The Cosmic Influence, F. King, Aldus Books, London, 1976.

+ Thanks are due to Dr. E. Mahers for Diagram 2.

** Others[3] have recommended the use of planets' rising points as an alternative to either the Ascendant or to the Morin Point. The problem with the rising point (mundane square to the M.C., i.e. measured along the equator)) is that it is individual for each planet concerned. Fot astrological purposes we require a point close to rising points through which essentially all of the planetary "influences" can act (with the possible exception of Pluto when at high celestial latitudes). In this way we can generate a full horoscope along with its subsequent, integrated interpretation. The Morin Point (zodiacal square to the M.C., i.e. measured along the ecliptic) provides one such simple point, along with its own astrological logic.

- -

CHAPTER 6

The Pre-Natal Epoch and the Ideal Birth Moment

Ancient teaching tells us that the Moon is the chief controller of human generation. Concerning our individual origins there are two moments, that of birth and that of fertilisation, that are important astrologically. The only astrological rules that are anywhere close to fitting the biological sequence of events described in Chapter 3 belong to the "Trutine of Hermes". These were developed primarily by Sepharial (W. Gorn-Old) and published in book form by Bailey[1] as "The Pre-Natal Epoch" in 1916. The rules contain four variables: i.e. the sex of the baby-to-be, and the positions of the Moon, of the Ascendant and, to a lesser extent, of that of the Sun. The only one of these about which we can have any reasonable doubt is that of the Ascendant. From experience there was difficulty using the Pre-Natal Epoch with the traditional Ascendant. Briefly there appeared to be far too many either short or long term births when compared to the usual distribution of gestation periods around 273 days. Using the Morin Point instead seemed to restore the expected distribution.

The rules state (but substituting the Morin Point for the Ascendant):
1) When the Moon at Birth is increasing in light (i.e. going from the new to the full Moon) it will be the Morin Point at Epoch, and the Moon at Epoch will be the Morin Point at Birth.
2) When the Moon at Birth is decreasing in light (i.e. going from the full to the new Moon) it will be the point opposite the Morin Point at Epoch, and the Moon at Epoch will be the point opposite the Morin Point at Birth.

These are extended by two further rules:

3) When the Moon at Birth is increasing in light and below the Morin Point, or when decreasing in light and above the Morin Point, then the period of gestation will be longer than the average.

4) When the Moon at Birth is increasing in light and above the Morin Point, or when decreasing in light and below the Morin Point, then the period of gestation will be shorter than the average.

From these four laws we can define Four Orders of Regular Epochs as follows in the Table:

Order	Condition	Period of Gestation
1	Moon above and increasing	273 days – x
2	Moon above and decreasing	273 days + x
3	Moon below and increasing	273 days + x
4	Moon below and decreasing	273 days – x

- -

273 days corresponds to the normal period of gestation (ten cycles of the Moon) counted backwards from the date of birth and gives the "Index date". x is the number of days equivalent to the number of degrees of the Moon with respect to the Morin Point, or to its opposite, at birth, depending on the Order number, divided by 13. (On average the Moon travels 13^0 per day around the ecliptic [Zodiac]). + or – x days from the Index date then gives the "Epoch date".

There then follow three sex rules governing the positions of the Moon and of the Morin Point (<u>both at Epoch</u>) that allow us to determine the sex of the baby-to-be (the Table of Bailey's Sex Degrees is given in Appendix 7): For the Epochal chart:

1) When the Morin Point is negative (non-sex), as in strict regular and irregular epochs, the sex of the position occupied by the Moon will be the sex of the baby-to-be.

2) When the Moon and the Morin Point positions are within orbs of degrees of the same sex, the sex of the baby-to-be is the same as the positions so occupied.

3) When the Moon and the Morin Point positions are placed within orbs of degrees of the opposite sex – the Moon in a female position

38

and the Morin Point in a male position, or <u>vice-versa</u> – the sex of the baby-to-be is determined by the sex of the quadrant containing the Moon. S.E. (Meridian to Morin Point) and N.W. quadrants are male and N.E. (Meridian to opposite to the Morin Point) and S.W. quadrants are female. The figure (Appendix 7) shows the sex influence of the quadrants at a glance. Notice how well they fit in with the Morinus House System.

We have seen already that the sex of the baby-to-be is decided at the moment of fertilisation (step 4) by the invading sperm. In practice, and almost without exception, we know the sex of the child and then we must find an Epoch that agrees with this.

The rules show that at the Pre-Natal Epoch (moment of fertilisation, step 4) there is a particular configuration of the heavens (mainly the Moon, but also the Sun) with respect to the Earth (Morin Point). It then follows, some nine months (ten cycles of the Moon) later, that there will be a related configuration of the heavens occurring at what we can call "The Ideal Birth Moment". Appendix 6 gives a full example.

One reason for having the ideal birth moment is to try to dispense with those influences at work to change the birth time so that it no longer occurs when it should. Major influences include induction by drugs, accidents (e.g. falling off a horse), the use of forceps, Caesarean births, breech births and the birth of a second, non-identical twin, but there must also be several, minor influences, all of which can lead to changed birth times. By contrast the moment of fertilisation (Pre-Natal Epoch) <u>in vivo</u> is highly sheltered, thermally constant and relatively uncomplicated compared with the process of birth, so that, potentially, it is a more reliable indicator of what the birth time should be. The problem here of course, is that we can't "see" when it happens.

Traditionally rectification is the process whereby we convert approximate or uncertain birth times into more accurate ones for correctly casting the horoscope. The use of the rules of the Pre-Natal Epoch for this purpose was claimed to be "the only reliable method of rectification". In the past astrologers regarded the moment of birth as sacrosanct, but realising how easily birth times can (be) change(d) perhaps we should regard the elaborate process of birth merely as a good indication of when a particular life on Earth begins, and that it simply is the best, so far, that nature can do. In other words a life should begin when a particular configuration of the heavens with respect to the Earth occurs, and that the actual birth moment, approximates, as closely as it can, to it. If we see the process of birth in this way, then rectification plays a far more important role, namely that of using the actual birth moment to determine what "the ideal birth moment" should be.

The idea of the Pre-Natal Epoch is a consequence of the genetic nature of the whole birth process. Thus Addey[7] has suggested, and recent results tend to confirm[21], that not only do we derive our physical form, but also our personalities from our genes, that have all become established and essentially unchangeable, at our moment of fertilisation. Now the interpretation of our horoscopes, cast for the dates, times and places of our moments of birth, also describes our personalities; i.e. the genetic traits already established at fertilisation must match those described at birth by interpretation of our horoscopes. This can only happen at one preordained time – the Ideal Birth Moment. We can conclude, despite the well-accepted view that our independent lives begin with our first breath, that for each and every one of us, there is a natural, or ideal, time to be born. This personal and particular time is an important birthright, which we lose whenever our birth times no longer occur when they should, e.g. for the reasons already given. Similarly, and topically, we become displaced from our "time" by cryoscopic preservation techniques, by the process of human cloning, as well as potentially "unfit" by indiscriminate injection of sperm into eggs.

40

Concerning the moments of birth and of fertilisation, <u>is it not possible that the Gauquelins' elite professionals, collectively showing enhanced planetary influences at birth, not only have favourable birth horoscopes, but that these are reinforced by favourable epochal ones? Is it then this combination that confers upon new-born-babies the potential status of elite professional?</u> At first sight one objection to this comes from the observed enhancement of planetary influences on geomagnetically disturbed days. But hard planetary aspects at the moment of birth are just the sort to lead to both geomagnetic disturbances <u>via</u> the solar wind from enhanced sunspot activity and the personality characteristics required for elite professionals. What we could have here then, is an example of true synchronicity.

We have now summarised Astrology's contribution to our birth on Earth. Our next steps (see Chapters 7 – 10) apply all of what we have presented to several given situations.

- -

References and footnotes: (lower numbered references occur at the ends of previous chapters.)

21a G. Turner of Hunter Genetics in The Lancet, 1996.

  b Lesch, Neils et al., Univ. Wurtzburg; Murphy of N.I.H., in Science, 1996.

  c Cloninger et al., Washington University in St. Louis, USA/Herzog Memorial Hospital, Jerusalem, Israel.

- -

CHAPTER 7

Forward from the Pre-Natal Epoch

"Know thee then that McDuff was from his mother's womb untimely
ripped."
Shakespeare, *Macbeth*

To the prospective father of two girl babies-to-be, each born
independently by necessary, elective, Caesarian section, the thought
was appalling that they should be born without any knowledge of
their natural birth time. This would mean that they would both be
deprived of a credible, natal horoscope so important for parents as a
valuable guide throughout their formative years. Of course, the times,
dates and places of their Caesarian births would be known, but in my
opinion these would not be acceptable for casting valid birth
horoscopes for the reasons given in the previous chapter. Was there
anything that could be done to identify the ideal birth times of the two
girl babies-to-be? It seemed that the only hope lay with the use of the
Pre-Natal Epoch[1]. Here we should need to determine the time of the
Epoch (i.e. of the moment of fertilisation) and from that derive an
Ideal Birth Time.

Fortunately recent medical knowledge allows us a much better
understanding of the events leading up to conception than was
available in Bailey's time. For example, we now know that most
women become fertile a fortnight or so before the onset of their next
menstruation. Additionally, we know that the average time from
insemination to fertilisation is seven hours with a wide distribution,
and that conception takes place some three or four days later. With
this in mind, times of artificial insemination[++] by husband were noted
month by month so that we should know, with hindsight, and with
some certainty, which one led to conception. Using this method, times

of insemination to fertilisation are likely to be shorter than the average. Thus we should know, within a few hours, the time of the moment of fertilisation. Moreover, three months later, following genetic tests (e.g. amniocentesis), we should know the sex of the baby-to-be.

For the Epochal horoscope we need to follow Bailey's sex rules given in the previous chapter:

1) The sex of the baby will be the sex of the position occupied by the Moon when the Morin Point is negative (non-sex).

2) The sex of the baby will be the same as that of the positions occupied by the Moon and by the Morin Point when these are the same and

3) The sex of the baby will be the same as that of the quadrant containing the Moon, when the sex of the positions held by the Moon, and by the Morin Point, are different

The first case

A successful insemination took place at 11:00 p.m. on February 7th, 1990 on the Wirral, England, presumably resulting in fertilisation several hours later, and conception a few days after that. Three months later genetic testing showed that the sex of the baby-to-be was female*.

Appendix 7 gives the detailed procedure that produced the three possible Epochs detailed in Table 1. Inspection of Table 1 shows that although the dates of birth are different, the Morin Points at birth are close to each other. This gives the three speculative horoscopes at birth a similar look (see Diagrams 1, 2 and 3). This is due to the Moon's position at Epoch varying little within the timescale of the three possible Epochs.

Table 1: Epochs early on the 8th February, 1990 (when the Sun's Position is around 18⁰ Aquarius) and their corresponding, Speculative Birth Data.

Epoch Number	Time of Epoch	Order and Type of Epoch	Morin Point at Epoch	Moon's Position at Epoch	Sex of Quadrant containing the Moon
1	00:21	1 Sex Irreg. 2nd Var.	17⁰47′ Scorpio	26⁰52′ Cancer	F
2	00:25	4 Sex Reg	18⁰35′ Scorpio	26⁰56′ Cancer	F
3	04:47	1 Sex Irreg. 2nd Var.	28⁰45′ Capricorn	29⁰22′ Cancer	F
Birth Number	Date of Birth	Time of Birth	Morin Point at Birth	Moon's Position at Birth	Sun's Position at Birth
1*	20 10 1990	11:57*	26⁰52′ Capricorn	17⁰47′ Scorpio	26⁰50′ Libra
2	03 11 1990	11:02	26⁰56′ Capricorn	18⁰35′ Taurus	10⁰46 Scorpio
3*	26 10 1990	11:42*	29⁰22′ Capricorn	28⁰45′ Capricorn	02⁰48′ Scorpio

*No allowance has been made here for British Summertime in operation.

Consulting Bailey's Table of Sex Degrees we see that the sex of the position of the Moon at Epoch is opposite to that of the baby-to-be, and so sex rule 3 is the one that applies. In this case there are not scores of acceptable Epochs, leading to an intractable morass of speculative birth times, but rather there appear to be surprisingly few of them. This is because the position of the Moon in the Zodiac at Epoch limits the number of opportunities on each potential birthday to two, i.e. when that Zodiac degree either rises or sets.

However, the positions of the Morin Point at Epoch, and hence the positions of the Moon at Birth, are substantially different (see Diagrams 1, 2 and 3), especially when the Order of the Epochs is different. Thus when trying to decide astrologically which of these Epochs is likely to be the correct one we shall examine the difference in the aspects that the Moon receives, and secondly, if necessary, in its position by both Sign and House, at Birth.

Diagrams 1, 2 & 3: The Three Speculative Birth Charts for the First Case.

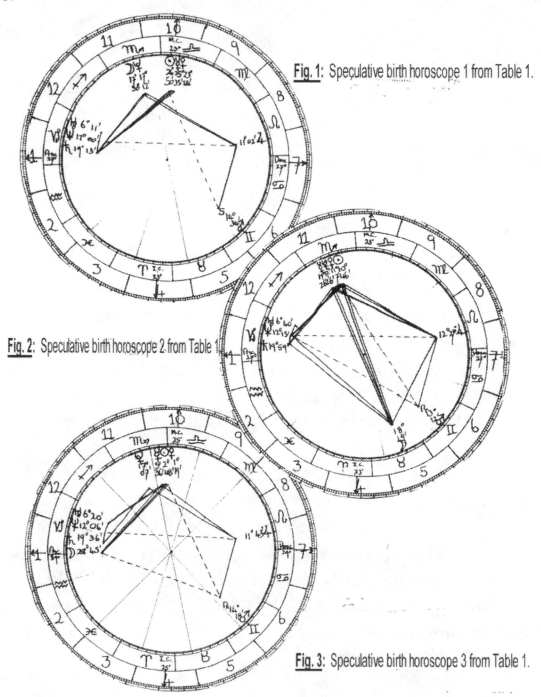

Fig. 1: Speculative birth horoscope 1 from Table 1.

Fig. 2: Speculative birth horoscope 2 from Table 1.

Fig. 3: Speculative birth horoscope 3 from Table 1.

As the characteristics ascribed to the Moon are strongly evident in young children[22], these differences should be clearly apparent at an early age. Table 2 enables us to compare the differences in the aspects that the Moon receives, and in the Moon's positions, for these three cases.

<u>Table 2</u>: The Moon's Position and the Aspects it Receives in Speculative Birth Horoscopes 1, 2 and 3.

Speculative Birth Horoscope Number	Sign Containing The Moon	House Containing The Moon	ASPECTS RECEIVED BY THE MOON FROM:- (ᴱ stands for exact)										
			Sun	Mer	Ven	Mar	Jup	Sat	Ura	Nep	Plu	MoPo	M.C.
1	Scorpio	11ᵗʰ.	—	—	—	—	□	✳	—	—	☌ᴱ	—	—
2	Taurus	5ᵗʰ.	☍	☍ᴱ	☍	□	△		△	☍ᴱ	—		
3	Caprico	1ˢᵗ.	□	□	□	□ᴱ	—	—	—	—	—	☌ᴱ	□

Table 2 shows that the differences between the three Moon positions, and the aspects that each receives, are strong and distinctive. Thus the Moon in possibility 1 receives no major aspects from the Sun, Mercury and Venus, but stands in close conjunction with Pluto. The Moon for possibility 2 receives aspects from most of the planets, in particular close oppositions to both Mercury and Pluto. The Moon in possibility 3 receives no major aspects from Jupiter, Saturn, Neptune and Pluto, but is conjoint the Morin Point.

My own experience and understanding of my daughter lead me to conclude that speculative birth horoscope 3 describes her best. Her Moon at the Morin Point gives her the soft constitution that she undoubtedly has. Incidentally, her mother also has her Moon at the Morin Point. Additionally, there is an exact sesquiquadrate to Mars that accounts for her somewhat feisty nature. Finally, this is the birth chart the Epoch of which has the most acceptable time between insemination and fertilisation, even though the time opening for a girl birth is short-lived (see Appendix 7).

We can summarise step-by-step, the whole process of what we have just described (as well as the content of Appendix 7 for the first case) as follows:

1) Find the sex of the position occupied by the Moon about the time of fertilisation. For now, and for simplicity, this needs to be opposite to that of the baby-to-be.

2) Find those periods about the time of fertilisation when both the sex of the position occupied by the Morin Point, and by the quadrant containing the Moon, are the same as that of the baby-to-be.

3) Decide, in order, possible dates of birth of those periods, by inspection of an Ephemeris, and determine when the Zodiacal longitude of the Epochal Moon either rises or sets on these days.

4) Now eliminate those possibilities where the Zodiacal longitude of the Moon at birth occupies a negative (non-sex) position. Such a position of the Moon at birth cannot produce a Morin Point at Epoch capable of countering the opposite sex of the position occupied by the Moon at Epoch (see also case 2).

5) For each remaining possibility determine the Order of the Epoch as well as the Index date, so that those that cannot be valid, based on the Index Date, are eliminated.

6) Eliminate those possibilities that cannot be valid because the sex of the quadrant containing the Moon is opposite to that of the baby-to-be.

7) Assemble all those valid Epochs left, together with their speculative birth times.

8) Cast their horoscopes and determine the most likely one, based largely on the differences between the aspects received by their natal Moons.

<u>The Second Case</u>.

Consideration of the second case should help to make the entire summary clearer. Once again a successful insemination took place at 08:30 a.m. on 28th April, 1986, on the Wirral, when British summertime was operating. An ultrasound scan, not long before the baby was born, showed that the baby-to-be was a girl[+]. Appendix 7 gives the detailed procedure used to produce the three possible Epochs found. Table 3, like Table 1, summarises the relevant data concerning them.

<u>Table 3</u>: Epochs on the Morning of the 28th April, 1986 (when the Sun's position is around 8^0 Taurus) for Girl Births, and their Corresponding, Speculative Birth Data.

Epoch Number	Time of Epoch*	Order and Type of Epoch	Morin Point at Epoch	Moon's Position at Epoch	Sex of Quadrant Containing the Moon.
1	09:12	3 Sex Regular	$03^045'$ Gemini	$00^045'$ Capricorn	F
2	09:56	3 Sex Regular	$15^044'$ Gemini	$01^015'$ Capricorn	F
3	10:38	3 Sex Regular	$27^030'$ Gemini	$01^045'$ Capricorn	F
Birth Number	**Date of Birth**	**Time of Birth**	**Morin Point at Birth**	**Moon's Position at Birth**	**Sun's Position at Birth**
1	07 02 1987	03:10	$00^045'$ Capricorn	$03^045'$ Gemini	$17^040'$ Aquarius
2	08 02 1987	03:06	$01^015'$ Capricorn	$15^044'$ Gemini	$18^040'$ Aquarius
3	09 02 1987	03:03	$01^045'$ Capricorn	$27^030'$ Gemini	$19^040'$ Aquarius

* Allowance has been made here for British Summertime in operation.

Examination of Table 3 shows that although the dates of birth are different, the Morin Points are closely similar, giving the three speculative horoscopes at birth a similar look. However, the positions of the Morin Point at Epoch, and hence the positions of the Moon at Birth, are significantly different. Note that, once again, the sex of the

48

Moon's position at Epoch is opposite to that of the baby-to-be, meaning that sex rule 3 is the one that applies here. Hence there are not scores of acceptable Epochs, but relatively few of them.

To decide astrologically which of these is the one most likely to be correct we shall examine once more, differences in the aspects that the Moon receives, and in its positions, at birth. Table 4 helps us to compare these differences:

<u>Table 4</u>: The Moon's Position and the Aspects it Receives in Speculative Birth Horoscopes 1, 2 and 3 in Table 3.

Speculative Birth Horoscope Number.	Sign Containing The Moon	House Containing The Moon	ASPECTS RECEIVED BY THE MOON:- (S stands for strong, M stands for medium)										
			Sun	Mer	Ven	Mar	Jup	Sat	Ura	Nep	Plu	MoPo	M.C.
1	Gemini	6th.	–	□ S	⊼	∠ –	–	–	–	–	–	–	△
2	Gemini	7th (just)	△ S –	–	–	–	–	☌ M	–	–	–	–	
3	Gemini	7th.	△ –	–	☌ –	–	□ S	–	☌ S –	–	□	☌	□

Table 4 shows, once again, that the differences between the aspects that the Moon receives are strongly significant. For possibility 1 the Moon experiences a strong square aspect from Mercury. In possibility 2 the Moon receives a strong trine from the Sun and a medium opposition from Saturn. But in possibility 3 there is a strong square from Jupiter, and a strong opposition from Uranus, to the Moon. Here also, experience and understanding of my daughter lead me to conclude that the interpretations for possibility 3 fit her character and behaviour extremely well. Additionally, this is the birth chart, the Epoch of which has the most acceptable time between insemination and fertilisation.

For the future, firstly we should extend the foregoing exercise to include cases of boys of uncertain birth time, whose Moon at Epoch occupies a female position. Perhaps then we could be ready to extend the method to cover all babies of uncertain birth time, regardless of their sex, and also of the sex of the position of the Moon at Epoch.

Normally birth is expected to occur 40 weeks after the first day of the last menstruation. For essentially half of elective Caesarian births (that are increasing these days because women are delaying having children until later in life), provided that we can identify a likely time for the moment of fertilisation, we should be able to estimate more accurately, using the procedures just described, when the natural birth would be likely to occur. This could prove useful for determining optimum times for carrying out Caesarian births, particularly when the gestation period is suggested to be significantly shorter than the average.

- -

References and footnotes: (lower numbered references occur at the ends of previous chapters.)

22) Planets in Youth, R. Hand, Whitford Press, Pennsylvania, USA, 1977, pp 103 – 142

* She was born by Caesarian Section at 11:15 a.m. at Arrowe Park Hospital, Wirral, England on the 18th October, 1990.

+ She was born by Caesarian Section at 10:42 a.m. at Arrowe Park Hospital, Wirral, England on the 15th January, 1987.

++ No account has been taken of the phenomenon of sperm capacitation. It has been assumed that all the times from insemination to fertilisation encountered in this chapter have been sufficient to accomplish this. Presumably this must be true.

- -

CHAPTER 8

ASTROGENEALOGY: Lines of Descent.

Both Astrology and Genealogy (Family History) start from the birth date, although Astrology works better when the birth time is also known. Birth dates slot into a time line for constructing family trees, that can be augmented from wills, local, national and social history, directories, service records, newspaper cuttings, surname studies, the Internet and especially photographs. But what is clearly lacking here, and what Astrology is so uniquely qualified to supply, is the personalities of the ancestors and how they interacted with each other. Probably we are far more interested in what our ancestors were like, rather than in the social conditions under which they lived. Take my wife's paternal, maternal, great grandmother, Harriet Gosset. She was born on the 2nd April 1841, in Hackney, London. She was descended from French Huguenots, and because the French record both birth times and birth dates we were fortunate to know that she was born around 02:00 a.m. Figure 1 shows her horoscope and aspect grid and Figure 2 shows a picture of her taken about 1910.

Fig 1: Horoscope of Harriet Gosset.   Fig 2: Photograph of Harriet Gosset.

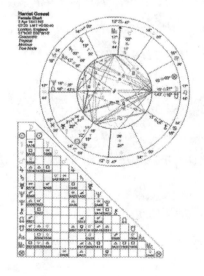

51

1) Person Summary

Figure 1 shows that her Morin Point occurs at 18⁰ Aquarius, her Sun is at 12⁰ Aries and her Moon lies close to that point opposite to the Morin Point at 210 Leo. Neptune is rising, Mars lies close to the Midheaven in Scorpio, and Uranus, her ruler, conjoins Mercury in Pisces.

Astrologers have their own ways of producing Person Summaries, but we can separate ours into four small sections, namely character, relationships, career and health, that together produce a workable whole.

Generous, good-natured, impulsive, and excitable, you value your sense of compassion, as well as having your own need for praise and attention. Your approach to life is gregarious, unconventional, original and objective. You view life from an intellectual stance, but need to beware of becoming too scathing of the emotional, and of the need to follow rules. You are a fast talker and like to play games that amuse your probing, quick-thinking mind. Highly intuitive with a fine imagination yet you fear anything hidden. Thus you are likely to inhibit and ignore your own inner urges, possibly even leading to paranoia.

Communication plays a key role in your life. You are drawn to unusual people but make an enthusiastic and fun-loving friend. Although you have difficulty expressing your feelings in relationships, you have a deeply felt need for just one-to-one. You need to receive attention from your loved one and like to travel with him. Relationships in your beautiful home need to be harmonious, and within it you find routine comforting and safe. You value your home, your family and its traditions deeply.

As an assertive and freedom loving individual you are a powerful person with leadership skills. Authoritarian, conservative and responsible you are quick to get to the point, enjoying challenge, coupled with an intense need for recognition in your work.

You recognise the need for cleanliness in your sexual relationships.

Harriet Gosset married Alfred Wing and had eight children. The last of these, a daughter, christened Alice Angelina, was born 7th December 1878, at Shenley, speculatively at 01:30 a.m. Figure 3 shows her horoscope plus aspect grid and Figure 4 shows a photograph of her taken in 1933.

2) Synastry[23]

Alice Wing (my wife's paternal grandmother) married Ernest Manktelow, also the last of eight children, in late January, 1906. Figure 5 shows his photograph taken in 1933, and Figure 6 shows his horoscope and grid.

Fig 3: Horoscope of Alice
Angeline Wing.

Fig 4: Photograph of Alice Angeline Wing.

Fig 5: Photograph of Ernest
Manktelow.

Fig 6: Horoscope of Ernest Manktelow.

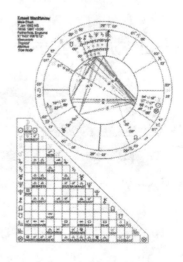

This time we shall perform a synastry operation using Ernest as the main recipient. Knowing what we know now, it may have been useful to say the following to Ernest at Christmas, 1905.

For Ernest and Alice true love is met, romance is set alight and there is a strong, sexual attraction that is combined with an enlivening, intellectual spark. Soul mates have been found and a strong and loving bond is formed: their relationship was written in the stars. Ernest is challenged to use his wits and there is fun, excitement and stimulation for them both. There is the thrill of being in each other's company, and although this may not last, they intuitively understand each other. A satisfying, long-term relationship results with mutual respect, as they feel able to tackle life's challenges together. This is because they complement each other when they join forces. Both begin to lead more creative lives.

Theirs is a life-changing relationship. Ernest feels life is more exciting and intense thanks to his relationship with Alice, but there is danger of a struggle for dominance. Nevertheless Alice shines through her association with Ernest and Ernest develops a great sense of well-being through his association with Alice. They enjoy picnics together in picturesque and romantic surroundings. They even concur on how money is saved, spent and enjoyed. Once obstacles have been overcome they embrace the happiness of the present. There is a happy balance between responsibility and diversity so that they grow more emotionally secure, allowing each other certain independence. They communicate comfortably and enjoy each other's company having so much in common. Overall they think that they have an ideal combination for an intimate relationship.

We can apply these techniques to all family tree members, given their birth data.

3) <u>Planetary Heredity</u>.

Perhaps a greater challenge for astrogenealogy consists of identifying inherited planetary characteristics from parents to children throughout a family tree. We know that personality traits are inherited, at least in part, hence a genuine planetary effect on personality might also be partly inherited[+]. A short review (see

Appendix 8) of the literature indicates that we can examine three likely sources of planetary heredity, namely: harmonic analysis in the Zodiac circle, Morinus sector analysis in the daily circle and by aspect analysis in the ecliptic (Zodiac) circle. These three analyses have been applied to three lines of descent:

1) <u>The Line of Gosset Descent</u>.

The members of my wife's side of the family through her father and back to his Gosset forefathers appear to possess a certain, genetically dominant, innate intelligence. To see if we can observe this astrologically, eight early recorded birth times of Gosset ancestors plus those of three direct descendants were examined. They were rectified to Ideal Birth Moments[1], Epochs noted and resulting horoscopes analysed thricely as indicated by the review. For comparison purposes a control group of eleven persons born at roughly the same time and place as the Gosset group members but related neither to themselves, nor to the Gosset group, was examined in the same way. Table 1 gives the details of the Gosset group:

<u>Table 1</u>: Horoscope details of the Gossets and of their Direct Descendants.

Gosset Member	Relationship to Harriet G	Time of Birth	The I.B.M.	Date of Birth	Place of Birth	Type of Epoch	Time and Date of Epoch
1	Great Aunt	09:00	08:57	23 9 1785	London	Order 2 Irreg. Sex	16:24 22 12 1784
2	Uncle	20:45	20:49	13 8 1808	London	Order 4 Reg. Sex.	04:59 25 11 1807
3	Aunt	04:10	04:12	2 10 1810	London	Order 3 Reg. Sex.	03:06 28 12 1809
4	Uncle	04:40	04:24	18 12 1812	London	Order 1 Reg. Sex.	11:06 1 4 1812
5	Aunt	02:00	02:14	25 9 1814	London	Order 1 Irreg. Sex.	06:27 22 1 1814
6	Herself	02:00	02:21	2 4 1841	London	Order 1 Irreg. Sex.	19:23 16 7 1840
7	Sister	07:00	06:40	13 2 1843	London	Order 3 Reg.	11:14 4 5 1842
8	Nephew	07:00	06:52	3 5 1865	East Grinstead	Order 3 Reg. Sex.	06:49 27 7 1864
9	Great Grand Daughter	B.S.T. 21:30	B.S.T. 21:31	30 8 1936	Bebington	Order 1 Irreg. 2.	09:13 19 12 1935
10	Great Grand Daughter	21:25	21:30	19 2 1946	Bromborough	Order 2 Irreg. Sex.	B.W.T. 18:06 9 5 1945
11	2X Great Granddaughter	B.S.T. 04:15	B.S.T. 04:09	15 9 1972	Bebington	Order 3 Irreg. Sex.	18:18 6 12 1971

General Aspect Analysis.

The total number of aspects (ca. 250) for both births and epochs to personal planets (i.e, Sun, Moon, Mercury, Venus and Mars) and to angles (Morin Point and Midheaven)[++] was about the same in both the Gosset and Control groups. Similarly the occurrence of minor aspects was roughly constant throughout. However, the most significant observation for Gosset group births was the lack of major aspects to the planet Venus (29 av. 42). The least common aspect was the trine and then the sextile (a lack of easy mental sharpness). This means that the Gosset group members tend to be awkward in their relationships with others. The next significant observation for Gosset group births was the number of major aspects to the Moon (54 av. 42) both from planets and to the Midheaven/Morin Point[++]. This means that the emotional side of their natures is well-integrated with the rest of their characteristics. These observations were noticeably not so apparent for the Control group.

Harmonic Analysis in the Zodiac (Ecliptic) Circle using Jigsaw V. 2 Astrological Computer Program[24].

An harmonic simply means the division of a circle of 360^0 by a whole number (integer).Thus the 1st harmonic is the division of 360^0 by 1, which gives exactly the same position on the circle again, i.e. an aspect of 0^0, or a conjunction. The 2nd harmonic divides the 360^0 by 2 giving 180^0, meaning that we have opposition aspects directly across the circle, and so on. The third harmonic gives a trine of 120^0; the 4th squares of 90^0 and the 6th sextiles of 60^0, etc. Harmonic analysis is more complete than aspect analysis and more flexible. The use of a computer saves an inordinate amount of time and effort for the researcher. (The orb of acceptance for the 1st harmonic was taken as + or - 8^0; subsequent orbs then become 8^0 divided by the number of the harmonic, e.g. the allowable orb for the 4th harmonic is thus + or - 2^0).

The 11 members of the Gosset group listed in Table 1, supplemented by 7 more having only approximate birth times, underwent harmonic analysis of both birth and epoch charts.

Similarly, 11 controls plus 7 more underwent the same analysis that involved 11 astrological planets (Sun, Moon etc. plus Chiron) together with an initial conjunction orb of 8^0. Only the first 12 harmonics were investigated. In these studies we were looking for harmonics present in the Gosset Group that were well-represented throughout the Group, but essentially lacking in the Controls.

Three harmonics stood out, i.e. the 11th, the 5th and the 10th in that order of significance ($\sum X^2$ = 23, 22 and 18) respectively. In particular the 11th harmonic is unusual but the interpretation given for this speaks of a group that *"seeks to drink from the Well of Knowledge. There is an active collective desire to gain some form of sacred or divine knowledge, which will give inspiration or clarity. This may be sought by studying but more likely by some kind of meditation or deepening process. The members of the group will know that what they seek cannot be gained by memorising figures and studying texts but rather by 'perambulating' the subject – winding around it, weaving in and out of it, like slowly understanding a great river by pausing often to drink from it, so that, little by little, you and the river become one."*

This interpretation could fit the Gossets quite well, from experience, and the interpretations of the 5th and 10th harmonics also seem close to what could have been expected.

Morinus Sector Analysis.

We are looking here at the daily circle where we start at the Morin Point and divide the whole circle into equal (36, 18 or 12) sectors. For Gosset Group births only sector 1 was significant (X^2 = 15). For Gosset Group epochs three sectors stood out (1, 7 and 2) but certainly not as strongly as sector 1 for the births*. Throughout the Control Group no outstanding sectors were apparent.

2) The Line of Male (Stubbs) Descent.

Here there were not enough known birth times to carry out analysis in the daily circle. Thus analysis was confined to the yearly Zodiac circle and to aspects.

Harmonic analysis of 40 separate, birth date charts resulted in little of significance save for a possible 7th harmonic, the interpretation of which could fit the Stubbs group well: *"The seventh harmonic has the potential both to inspire and ensnare. This family group has an emotional intensity, a touch of neurosis, and can hover on the edge of obsession, from which precarious position it can gain inspiration. Group members will both love and hate each other but if they can bind together for a while using psychic and spiritual links, they can become truly inspiring. Passionate, creative and emotionally unstable day to day business of the group often proves very difficult."*

Analysis by aspects of personal planets revealed an excess of conjunctions plus a possible excess of trines. This means that the group possesses assertive qualities dependent on the components of the conjunctions with little regard for the effect on others, but with some mitigation (trines).

3) <u>The Line of Female Descent</u>.

Here again there were insufficient known birth times to carry out analysis in the daily circle, but a pilot study of 8 charts using twelve sectors suggested an emphasis on the 1st sector. Harmonic analysis of 23 separate, birth date charts resulted in little of significance apart perhaps from a 1st harmonic.

Analysis by aspects of personal planets again showed little of significance except possibly for a small excess of conjunctions. Considering the importance of the line of female descent (see Ch. 3), connecting us with our ancestors, these results were disappointing.

- -

References and footnotes: (lower numbered references occur at the ends of previous chapters.)

23) Synastry, R. C. Davison, ASI Publications, New York, 1977. (Solar Fire v.6 can carry out synastry operations.)

24) Jigsaw v. 2 (developed by Bernadette Brady), Astrolabe Inc., Brewster, Maryland, USA.

*   Sometimes there can be three valid epochs for one particular birth. The difficulty then is to try to pick the one that seems most likely. For example the one with the more normal gestation period, or the one that contains more or less of the required sex degrees, or the one occurring around Christmas, New Year or birthday, may be preferred. Of course there is no guarantee that we have the correct one.

+   Unfortunately the same planet in different signs, houses and experiencing different aspects can nevertheless lead to similar interpretations, thereby complicating a situation that at first sight would seem to be straightforward. This has been mentioned simply as a precaution.

++   In this analysis major aspects to the Morin Point and to the M.C. were included because they have an identifiable impression on interpretation.

- -

CHAPTER 9

ASTROGENEALOGY: Single generation studies.

Since, in all probability, our genes (DNA) give us our characteristics, and since our horoscopes do just the same, then we have here two descriptions of one and the same thing. Thus if we find horoscope interpretations that describe the shared characteristics of a group, then presumably that group shares the same type of genes.

Generally siblings share 50% (1/2) of each other's DNA, although each 50% share is different from every other 50% share within the group. For Astrogenealogy surely we should expect to find some sort of group relationships existing among the horoscopes of large families.

<u>King George III's Children</u>

<u>Diagram 1</u>: King George III, Queen Charlotte and their first six children

King George III and Queen Charlotte had 15 children all by single births. Birth dates for all fifteen were readily available[2] and, fortunately, birth times[25], although not that accurate, were obtainable from local newspapers of the time. All the children were conceived and born in London. Table 1 gives the birth and epoch data (using the Ideal Birth Time and the Morinus House System[1]) for King George III's (KGIII's) children.

Table 1: Birth and Epoch Dates and Times for King George III's Children:

Name	Birth Date	Recorded Birth Time	Ideal Birth Time	Epoch Date	Epoch Time	Order and Type of Epoch
George	12 08 1762	07:30	07:13:32	07 11 1761	05:51:56	2 Reg. Sex.
Frederick	16 08 1763	10:00	09:48:40	31 10 1762	07:25:20	3 Irreg. 2nd Var.
William	21 08 1765	03:45	03:46:56	14 11 1764	16:00:04	3 Sex Irreg. 1st V.
Charlotte	29 09 1766	06:00-07:00	06:52:04	08 12 1765	09:33:04	2 Irreg. 2nd Var.
Edward	02 11 1767	12:00	11:56:36	30 01 1767	08:39:24	3 Sex Reg.
Augusta	08 11 1768	20:30	20:17:56	13 02 1768	11:14:16	4 Sex Reg.
Elizabeth	22 05 1770	08:00-09:00	08:15:19	13 08 1769	10:17:40	2 Sex Reg.
Ernest	05 06 1771	05:50	05:39:48	28 08 1770	06:11:36	2 Reg.
Augustus	27 01 1773	06:00	05:40:40	24 04 1772	15:41:00	3 Sex Irreg. 1st V.
Adolphus	24 02 1774	18:15	18:09:36	29 05 1773	10:25:04	1 Reg.
Mary	25 04 1776	07:00	06:39:04	20 07 1775	18:05:08	3 Irreg. 1st Var.
Sophia	03 11 1777	21:00	20:44:28	20 01 1777	15:21:56	3 Sex Irreg. 1st V.
Octavius	23 02 1779	03:00-04:00	03:18:32	02 05 1778	07:07:22	3 Sex Irreg. 2nd V
Alfred	22 09 1780	10:00	09:40:04	20 12 1779	18:32:12	2 Sex Irreg. 1st V.
Amelia	07 08 1783	02:30	02:30:28	25 10 1782	19:45:48	3 Sex Irreg. 1st V.

Analysis in the Daily Circle using Morinus Sectors.

Horoscopes for KGIII's children were cast using the ideal Birth Times in Table 1 and the Morinus House System[1]. The positions of the 11 planets (Sun, Moon etc. plus Chiron) for all of them, with respect to the Morin Point, were placed in one of 18 Morinus Sectors. Polar Graph 1 shows the result of plotting planet occupancy for each of the 18 sectors for both the births and epochs of KGIII's children. Because there is nothing that is significantly outstanding for either of these graphs (the distribution for both is relatively general) there was no need to run control experiments.

Polar Graph 1: The Sector Position (1 – 18) of all the Planets in the Horoscopes of KGIII's Children at Birth and at Epoch.

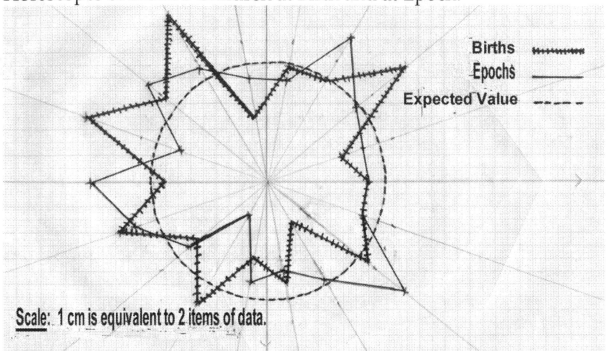

Births ┼┼┼┼┼┼┼┼┼
Epochs ——————
Expected Value — — — —

Scale: 1 cm is equivalent to 2 items of data.

Analysis by Aspects.

There was no favoured personal planet, or otherwise, or major aspect, in the birth horoscopes of KGIII's children. However, the epoch horoscopes showed a distinct lack of aspects to Mercury (supported somewhat by the birth horoscopes) and suggests that their mentality is not well-integrated with the rest of their characteristics. Additionally there was a noticeable excess of trines for the epoch group that perhaps lends support to the statement that King George III and Queen Charlotte had "such lovely children".

Analysis in the Ecliptic (Zodiac) Circle.

Analysis of all the planets of the horoscopes of KGIII's from their Ideal Birth Times and from their epochs, for outstanding harmonics from 1 – 12 only, was accomplished using the Jigsaw 2 program[24]. As the time range of the study only covers some 22 years (1761 – 1783), and because the movements of the generation planets (Jupiter – Pluto plus Chiron) tend to be slow, duplication (due mainly to apparent retrograde motion of the planets) often occurred. When this happened only one was counted. For the births, the 7th harmonic ($X^2 = 8.7$) is worth mentioning, with essentially nothing of significance for the epochs (6th; $X^2 = 4.6$).

From this single generation study we have seen that only aspect analysis of the epochs yielded significant results that appeared to fit for the children. This result does provide some support for the whole idea of epochs.

<u>Queen Victoria's Children.</u>

We have stated that, in general, siblings share 50% of each other's DNA. For Queen Victoria's (QV's) children that share increases to 62.5% (⅝) because she married her 1st cousin, Prince Albert, who was her mother's brother's son. Queen Victoria and Prince Albert had nine children, the birth times of each being carefully recorded, and available to the public[26].

<u>Diagram 2</u>: Queen Victoria with her First Five Children

Surely we should expect to find some sort of relationships existing among the horoscopes of these nine children. Table 2 gives the birth and epoch data for QV's children, who were all conceived and born in London.

Table 2: Birth and Epoch[1] Dates and Times for Queen Victoria's Children.

Name	Birth Date	Recorded Birth Time	Ideal Birth Time	Epoch Date	Epoch Time	Order and Type of Epoch
Victoria	22 11 1840	14:00	13:52:16	06 02 1840	23:59:36	2 Irreg. 1st Var.
Edward	09 11 1841	10:48	10:34:28	04 02 1841	20:58:12	2 Sex Irreg. 1st Var.
Alice	25 04 1843	04:05	03:51:52	13 07 1842	09:31:28	2 Sex Reg.
Alfred	06 08 1844	07:50	07:56:00	02 11 1843	06:29:48	2 Reg.
Helena	25 05 1846	14:55	14:42:48	04 09 1845	11:48:28	1 Irreg. 1st Var.
Louise	18 03 1848	08:00	07:56:00	26 05 1847	12:22:40	3 Reg.
Arthur	01 05 1850	08:17	08:16:44	30 07 1849	03:54:57	2 Reg.
Leopold	07 04 1853	13:30	13:28:34	21 06 1852	00:34:16	2 Irreg. 3rd Var.
Beatrice	14 04 1857	13:45	13:42:00	19 07 1856	15:37:08	4 Sex Irreg. 1st Var.

Analysis in the Daily Circle using Morinus Sectors.

The positions of all of the planets in the horoscopes of QV's children, with respect to the Morin Point, were determined and placed in one of 12 Morinus Sectors. Polar Graph 2 shows the results of plotting planet occupancy for each of 12 sectors for their births. The graph also shows the result of plotting planet occupancy for each of the 12 Morinus Sectors for the controls.

Surprisingly both results appear to be significant* ($\sum X^2 = 29.1$; p = 0.2) for QV's children and ($\sum X^2 = 24.8$; p = 1) differently, for the controls. Interestingly we can rationalise these results using M. E. Jones's interpretations[27]. The graph for QV's children shows a clear South and West emphasis. The interpretations here are *"objective"* and *"they have their destiny imposed upon them"*, both of which make sense for QV's children as a group. Here shared character of the family group corresponds with shared genes. Similarly the graph for the controls shows a strong East emphasis for which the interpretation

Polar Graph 2: The Sector Position (1 – 12) of all the Planets in the Birth Horoscopes of QV's Children (Plus Controls).

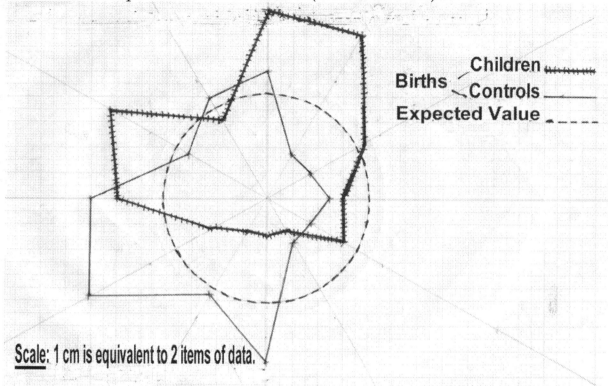

Births — Children
Controls
Expected Value

Scale: 1 cm is equivalent to 2 items of data.

is *"they have their destiny in their own hands"*. As the members of this group are all notables, for good or ill, this interpretation fits them well as a group. Once again shared character equates to shared genes, although not in a family sense as we have just seen, but rather as *"destiny deciders"*. If we accept these results then perhaps we need to impose an added constraint on controls to those of similar time and place in this kind of study – namely that all the planets in the horoscopes need to be well-dispersed through the sectors, or that there needs to be an even distribution overall. This added constraint could prove quite difficult to satisfy!

Similarly Polar Graph 3 shows the results of plotting planet occupancy for each of 12 Morinus Sectors for the epochs of QV's children, as well as for the epochs of the controls.

Polar Graph 3: The Sector Position (1 – 12) of all the Planets in the Epoch Horoscopes of QV's Children (Plus Controls).

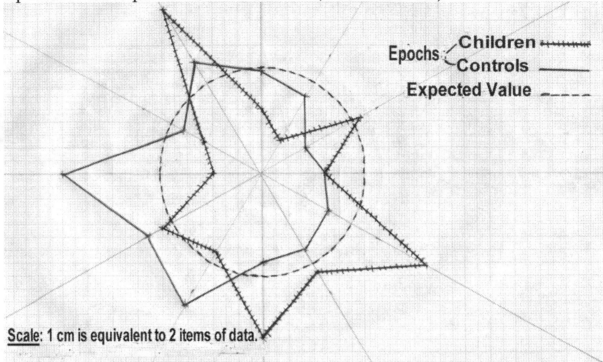

Epochs: Children ↔↔↔↔↔
Controls ————
Expected Value – – – – –

Scale: 1 cm is equivalent to 2 items of data.

Once again both results have some significance ($\sum X^2 = 23.5$; $p = 1.5$) for QV's children and ($\sum X^2 = 14.5$; $p = 20$) for the controls. Although it is difficult to rationalise the graph shape for the epochs of QV's children, it is nevertheless distinctive and so noteworthy. A small East emphasis shows up for the epochs of the controls. Perhaps true notability requires East emphasis for both birth and epoch horoscopes (see underlined section at the end of chapter 6)!

Analysis by Aspects.

There appeared to be no particularly favoured planet or otherwise for the birth or epoch horoscopes of QV's children. However, for the births there is a distinct lack of oppositions (a lack of tenseness) among the major aspects (as could have been deduced from Polar graph 3), as well as, to a lesser extent, an excess of conjunctions+.

Analysis in the Ecliptic (Zodiac) Circle.

Analysis of the horoscopes of QV's children from their Ideal birth and epoch times, for outstanding harmonics from 1 – 12, was effected using the Jigsaw 2 program. For background "noise" purposes

66

analyses were also carried out for control horoscopes using their Ideal birth and epoch times. However the time range for the whole study only spans some 17 years (1840 – 1857). During this period the movements of the generation planets tend to be slow so that when duplication occurred it was allowed for only once. After correcting for background noise using the controls, only the 1st harmonic for the births ($X^2 = 4.8$) is just worth mentioning. Again, for the epochs, only the 1st harmonic (two of them [$X^2 = 8.1$; 110^0 44' and 7.9; 137^0 39']) just about stand out.

Comparing the results of QV's children with those of KGIII's children in the daily circle we can wonder if the differences observed are in any way due to the different levels of shared DNA among the respective siblings. However, the results for the controls tend to counter this. From the QV's children study we have seen that only analysis in the daily circle yields particularly noteworthy results and so provides some support for astrogenealogy. Once again the fact that we observe significant results for epochs in the daily and ecliptic circles lends support for the proposal that the Morinus House System constitutes "The House System of Choice". The Epoch results also support the general idea of "The Pre-Natal Epoch".

Is it possible that taking all the astrogenealogical results together, we are starting to find evidence for the existence of a detectable link between the heavenly bodies of the Solar System and human life here on Earth?

- -

References and footnotes: (lower numbered references occur at the ends of previous chapters.)

25) Birth times were obtained by courtesy of the Royal Archives using the Annual Register. However the birth times of Prince Alfred and Princess Amelia were obtained by courtesy of the British Library (Newspapers) from "The Morning Chronicle" the day after each birth.

26) The Astrological Association's database (London) has all the birth details of Queen Victoria's immediate family.

* Significances throughout are comparative only. The assumption has been made that the planets of the Solar System can become distributed regularly around the daily circle. We know that is not true.

27) "How to Learn Astrology", M. E. Jones, Routledge and Kegan Paul, London, 1977, Ch.1.

CHAPTER 10

The Moon Alone

JULIET to ROMEO:
> "O! Swear not by the Moon, the inconstant Moon,
> That monthly changes in her circled orb,
> Lest that thy love prove likewise variable."

Shakespeare, *Romeo and Juliet*

We saw at the end of Chapter 8 that the results obtained for the line of female descent were disappointing. This was also puzzling because expectation here was highest. Up until now everything had gone favourably. Had something fundamental been missed?

Because there was faith in the Pre-Natal Epoch, which deals mostly with the Moon, the positions of the Moon both at birth and at epoch, with respect to the Morin Point, in the daily circle, were examined in those cases where there was confidence in the birth times and these were extrapolated to include the remainder as far as possible. The Polar Graph shows the results obtained.

Examination of the Graph showed firstly that the positions of the Moon at birth and at epoch for each case were reflected across one of two imaginary plane mirrors at right angles (in the ecliptic) in the horoscope as described in Chapter 1. It then became clear that this observation was true for every horoscope that had been cast using The Ideal Birth Time based on its appropriate epoch time[1]. Presumably this then is a consequence of carrying out the procedure of the Pre-Natal Epoch. However it is very tempting to propose that this is how animals and then humans managed to regulate their reproduction by the Moon in the first place.

<u>Polar Graph</u>: Sector Positions (1 – 12) of the Moon in the Horoscopes of the Line of Female Descent Group.

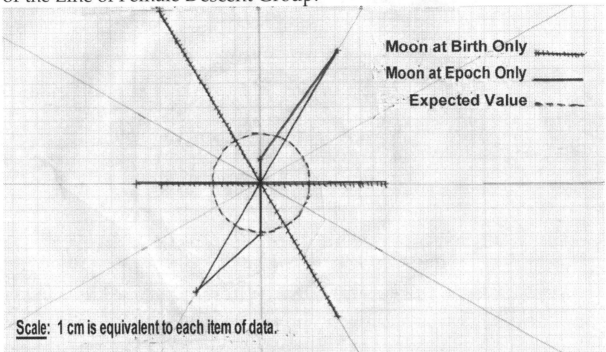

Moon at Birth Only _____

Moon at Epoch Only _____

Expected Value _ _ _ _ _

<u>Scale</u>: 1 cm is equivalent to each item of data.

Examination of the Graph of this Pilot Study also indicates that members of the group of the line of female descent, whose epochs were of the same order, had their Moon positions at birth at one or two of the same positions with respect to the Morin Point. For example when the Order of Epoch was 3 (Moon below and increasing at birth) then the angle from the Moon at birth, with respect to the Morin Point, was either about 175^0 or 125^0. Similarly when the Order of Epoch was 2 (Moon above and decreasing at birth) the angle was about 5^0 or 100^0. For Order 1 the angles were 5^0 or 60^0. Should this type of observation hold for other lines of female descent then a question that follows at once is: "Do the positions in the horoscopes of the remaining planets of the Solar System occur solely as their motions dictate, or are there other influences at work of which we are as yet unaware?" (e.g. see Ch. 9, Analysis by Aspects of KGIII's Children). Perhaps we are, after all, at least to a certain extent, creatures of chance. . . . . . .

- -

APPENDIX 5a

A Horoscope from First Principles

Consider, as an example, the horoscope of Prince Harry:

<u>Diagram 1:</u> The Horoscope of Prince Harry.

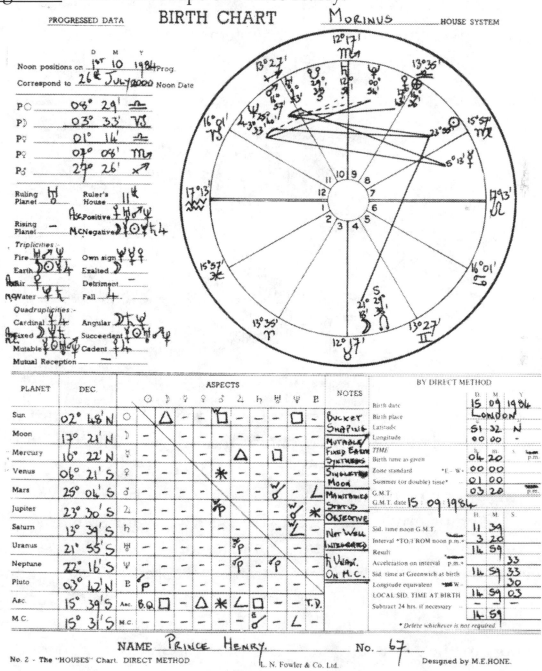

Looking at the column on the right, at the bottom (BY DIRECT METHOD) we see, by working our way down, that Prince Harry was born on the 15th September, 1984, in London that has a latitude 51⁰N32′ and a longitude that is essentially 0⁰W. He was born at twenty past four in the afternoon when British Summertime was operating so we subtract one hour to give Greenwich Mean Time of 03:20 p.m. At this stage we need to convert Sun time to sidereal (Star) time. This is done more conveniently by consulting an ephemeris.

Diagram 2 gives the ephemeris for September, 1984[28].

LONGITUDE **SEPTEMBER 1984**

DAY	SID.TIME	☉	☽	TRUE ☊	☿	♀	♂	♃	♄	♅	♆	♇
	h m s	° ' "	° ' "	° '	° '	° '	° '	° '	° '	° '	° '	° '
1 Sa	22 41 21	8♍ 44 3	20♏ 31 33	1♊D17.9	29♍R33.6	29♍ 44.3	7♐ 56.4	3♑ 8.0	11♏ 38.0	9♐ 36.8	28♐R40.7	0♏ 6.5
2 Su	22 45 18	9 42 8	4♐ 14 59	1R 17.8	1 51.1	0♎ 58.0	8 31.9	3 8.4	11 42.5	9 37.6	28 40.4	0 8.2
3 M	22 49 14	10 40 15	17 35 10	1 17.9	1 14.5	2 11.7	9 7.7	3 9.1	11 47.1	9 38.4	28 40.2	0 9.9
4 Tu	22 53 11	11 38 22	0♑ 35 0	1 16.9	0 44.5	3 25.4	9 43.7	3 10.0	11 51.7	9 39.2	28 40.0	0 11.6
5 W	22 57 7	12 36 32	13 17 50	1 13.7	0 22.1	4 39.1	10 19.9	3 11.0	11 56.4	9 40.1	28 39.8	0 13.4
6 Th	23 1 4	13 34 42	25 46 57	1 7.8	0 7.8	5 52.8	10 56.3	3 12.3	12 1.2	9 41.0	28 39.7	0 15.2
7 F	23 5 1	14 32 54	8♒ 5 13	0 59.1	0D 1.9	7 6.5	11 32.9	3 13.7	12 6.0	9 42.0	28 39.6	0 17.0
8 Sa	23 8 57	15 31 8	20 14 57	0 47.8	0 4.9	8 20.1	12 9.7	3 15.3	12 11.0	9 43.1	28 39.5	0 18.9
9 Su	23 12 54	16 29 23	2♓ 17 58	0 34.9	0 16.8	9 33.8	12 46.7	3 17.1	12 15.9	9 44.2	28D 39.4	0 20.8
10 M	23 16 50	17 27 41	14 15 41	0 21.4	0 37.7	10 47.4	13 23.9	3 19.1	12 21.0	9 45.3	28 39.4	0 22.6
11 Tu	23 20 47	18 25 59	26 9 24	0 8.6	1 7.4	12 1.1	14 1.2	3 21.3	12 26.1	9 46.5	28 39.4	0 24.6
12 W	23 24 43	19 24 20	8♈ 0 30	29♉ 57.3	1 45.8	13 14.7	14 38.8	3 23.7	12 31.3	9 47.7	28 39.5	0 26.5
13 Th	23 28 40	20 22 43	19 50 46	29 48.4	2 32.4	14 28.3	15 16.6	3 26.2	12 36.5	9 49.0	28 39.6	0 28.4
14 F	23 32 36	21 21 7	1♉ 42 30	29 42.3	3 27.1	15 41.9	15 54.5	3 28.9	12 41.8	9 50.4	28 39.7	0 30.4
15 Sa	23 36 33	22 19 34	13 38 40	29 38.9	4 29.1	16 55.4	16 32.7	3 31.8	12 47.2	9 51.8	28 39.8	0 32.4
16 Su	23 40 30	23 18 3	25 42 59	29D 37.7	5 38.1	18 9.0	17 11.0	3 34.9	12 52.6	9 53.2	28 40.0	0 34.4
17 M	23 44 26	24 16 34	7♊ 59 43	29 37.9	6 53.3	19 22.6	17 49.4	3 38.2	12 58.1	9 54.7	28 40.2	0 36.5
18 Tu	23 48 23	25 15 7	20 33 37	29R 38.4	8 14.4	20 36.1	18 28.1	3 41.6	13 3.6	9 56.2	28 40.5	0 38.5
19 W	23 52 19	26 13 43	3♋ 29 24	29 38.2	9 40.4	21 49.7	19 6.9	3 45.2	13 9.2	9 57.8	28 40.8	0 40.6
20 Th	23 56 16	27 12 20	16 51 13	29 36.4	11 11.1	23 3.2	19 45.9	3 49.0	13 14.9	9 59.4	28 41.1	0 42.7
21 F	0 0 12	28 11 0	0♌ 41 51	29 32.3	12 45.6	24 16.7	20 25.1	3 53.0	13 20.6	10 1.1	28 41.4	0 44.8
22 Sa	0 4 9	29 9 42	15 1 44	29 26.2	14 23.6	25 30.3	21 4.4	3 57.2	13 26.4	10 2.8	28 41.8	0 46.9
23 Su	0 8 5	0♎ 8 26	29 48 9	29 18.3	16 4.2	26 43.8	21 43.9	4 1.5	13 32.2	10 4.6	28 42.2	0 49.1
24 M	0 12 2	1 7 12	14♍ 54 46	29 9.5	17 47.2	27 57.3	22 23.6	4 6.0	13 38.1	10 6.4	28 42.7	0 51.2
25 Tu	0 15 59	2 6 1	0♎ 12 9	29 1.1	19 32.0	29 10.7	23 3.4	4 10.7	13 44.0	10 8.3	28 43.2	0 53.4
26 W	0 19 55	3 4 51	15 29 10	28 53.8	21 18.2	0♏ 24.2	23 43.3	4 15.5	13 50.0	10 10.2	28 43.7	0 55.6
27 Th	0 23 52	4 3 43	0♏ 34 58	28 48.6	23 5.5	1 37.7	24 23.5	4 20.5	13 56.0	10 12.1	28 44.2	0 57.8
28 F	0 27 48	5 2 38	15 20 51	28 45.7	24 53.5	2 51.1	25 3.7	4 25.7	14 2.1	10 14.1	28 44.8	1 0.0
29 Sa	0 31 45	6 1 34	29 41 18	28 44.9	26 41.9	4 4.6	25 44.2	4 31.0	14 8.2	10 16.2	28 45.4	1 2.2
30 Su	0 35 41	7♎ 0 32	13♐ 34 10	28♉ 45.4	28♍ 30.6	3♏ 18.0	26♐ 24.7	4♑ 36.5	14♏ 14.4	10♐ 18.3	28♐ 46.1	1♏ 4.5

We see that the sidereal (star) time for midnight at the start of the 15th September is 23:36:33. To obtain the sidereal time for noon we add 12hrs. 2mins. to this to obtain 35:38:33. Taking away 24 hrs and rounding up, gives us 11:39. To this we add the interval from noon, i.e. 3hrs. 20mins. giving us 14:59. Because we want star time rather than Sun time we need to add 3 and a third times 9.86 secs. (the correction factor for each hour of Sun time) = 33secs. Because London is slightly to the West of Greenwich there is a small longitude correction to subtract, that just about cancels out the interval correction in this, but only in this, case. Having obtained our local sidereal (star) time at birth we now look up a star time of 14:59 in Tables of Morinus Houses (see following) to find out the Morin Point in degrees of the Zodiac along with the values for the other House centres (underlined).

TABLES OF MORINUS HOUSES

S.T.	1	2	3	4	5	6
0 0	0 ♋ 0	2 ♌ 11	2 ♍ 5	0 ♎ 0	27 ♎ 55	27 ♏ 49
0 4	1 5	3 13	3 3	0 55	28 53	28 52
0 8	2 11	4 16	4 0	1 50	29 49	29 54
0 12	3 16	5 18	4 57	2 45	0 ♏ 47	0 ♐ 57
0 16	4 22	6 20	5 54	3 40	1 45	2 0
0 20	5 27	7 21	6 50	4 35	2 43	3 3
0 24	6 32	8 23	7 47	5 31	3 41	4 7
0 28	7 37	9 24	8 43	6 26	4 39	5 10
0 32	8 43	10 25	9 40	7 21	5 38	6 14
0 36	9 48	11 26	10 36	8 16	6 36	7 18
0 40	10 53	12 27	11 33	9 11	7 35	8 22
0 44	11 58	13 27	12 28	10 7	8 34	9 26
0 48	13 3	14 28	13 24	11 2	9 33	10 30
0 52	14 7	15 28	14 20	11 57	10 33	11 34
0 56	15 12	16 28	16 16	12 53	11 32	12 39
1 0	16 17	17 28	16 11	13 49	12 32	13 43
1 4	17 21	18 28	17 7	14 44	13 32	14 48
1 8	18 26	19 27	18 3	15 40	14 32	15 52
1 12	19 30	20 27	18 58	16 36	15 32	16 57
1 16	20 34	21 26	19 53	17 32	16 33	18 2
1 20	21 38	22 25	20 49	18 27	17 33	19 7
1 24	22 42	23 24	21 43	19 24	18 34	20 12
1 28	23 46	24 22	22 39	20 20	19 35	21 17
1 32	24 50	25 21	23 34	21 17	20 36	22 23
1 36	25 53	26 19	24 29	22 13	21 37	23 28
1 40	26 57	27 17	25 25	23 10	22 39	24 33
1 44	28 0	28 15	26 20	24 6	23 40	25 38
1 48	29 3	29 13	27 15	25 3	24 42	26 44
1 52	0 ♌ 6	0 ♍ 11	28 10	26 0	25 44	27 49
1 56	1 8	1 7	29 5	26 57	26 47	28 55
2 0	2 ♌ 11	2 ♍ 5	0 ♎ 0	27 ♎ 55	27 ♏ 49	0 ♑ 0
2 4	3 13	3 3	0 55	28 53	28 52	1 5
2 8	4 16	4 0	1 50	29 49	29 54	2 11
2 12	5 18	4 57	2 45	0 ♏ 47	0 ♐ 57	3 16
2 16	6 20	5 54	3 40	1 45	2 0	4 22
2 20	7 21	6 50	4 35	2 43	3 3	5 27
2 24	8 23	7 47	5 31	3 41	4 7	6 32
2 28	9 24	8 43	6 26	4 39	5 10	7 37
2 32	10 25	9 40	7 21	5 38	6 14	8 43
2 36	11 26	10 36	8 16	6 36	7 18	9 48
2 40	12 27	11 33	9 11	7 35	8 22	10 53
2 44	13 27	12 28	10 7	8 34	9 26	11 58
2 48	14 28	13 24	11 2	9 33	10 30	13 3
2 52	15 28	14 20	11 57	10 33	11 34	14 7
2 56	16 28	15 16	12 53	11 32	12 39	15 12
3 0	17 28	16 11	13 49	12 32	13 43	16 17
3 4	18 28	17 7	14 44	13 32	14 48	17 21
3 8	19 27	18 3	15 40	14 32	15 52	18 26
3 12	20 27	18 58	16 36	15 32	16 57	19 30
3 16	21 26	19 53	17 32	16 33	18 2	20 34
3 20	22 25	20 49	18 27	17 33	19 7	21 38
3 24	23 24	21 43	19 24	18 34	20 12	22 42
3 28	24 22	22 39	20 20	19 35	21 17	23 46
3 32	25 21	23 34	21 17	20 36	22 23	24 50
3 36	26 19	24 29	22 13	21 37	23 28	25 53
3 40	27 17	25 25	23 10	22 39	24 33	26 57
3 44	28 15	26 20	24 6	23 40	25 38	28 0
3 48	29 13	27 15	25 3	24 42	26 44	29 3
3 52	0 ♍ 11	28 10	26 0	25 44	27 49	0 ♒ 6
3 56	1 7	29 5	26 57	26 47	28 55	1 8
4 0	2 ♍ 5	0 ♎ 0	27 ♎ 55	27 ♏ 49	0 ♑ 0	2 ♒ 11
4 4	3 3	0 55	28 53	28 52	1 5	3 13
4 8	4 0	1 50	29 49	29 54	2 11	4 16
4 12	4 57	2 45	0 ♏ 47	0 ♐ 57	3 16	5 18
4 16	5 54	3 40	1 45	2 0	4 22	6 20
4 20	6 50	4 35	2 43	3 3	5 27	7 21
4 24	7 47	5 31	3 41	4 7	6 32	8 23
4 28	8 43	6 26	4 39	5 10	7 37	9 24
4 32	9 40	7 21	5 38	6 14	8 43	10 25
4 36	10 36	8 16	6 36	7 18	9 48	11 26
4 40	11 33	9 11	7 35	8 22	10 53	12 27
4 44	12 28	10 7	8 34	9 26	11 58	13 27
4 48	13 24	11 2	9 33	10 30	13 3	14 28
4 52	14 20	11 57	10 33	11 34	14 7	15 28
4 56	15 16	12 53	11 32	12 39	15 12	16 28
5 0	16 11	13 49	12 32	13 43	16 17	17 28
5 4	17 7	14 44	13 32	14 48	17 21	18 28
5 8	18 3	15 40	14 32	15 52	18 26	19 27
5 12	18 58	16 36	15 32	16 57	19 30	20 27
5 16	19 53	17 32	16 33	18 2	20 34	21 26
5 20	20 49	18 27	17 33	19 7	21 38	22 25
5 24	21 43	19 24	18 34	20 12	22 42	23 24
5 28	22 39	20 20	19 35	21 17	23 46	24 22
5 32	23 34	21 17	20 36	22 23	24 50	25 21
5 36	24 29	22 13	21 37	23 28	25 53	26 19
5 40	25 25	23 10	22 39	24 33	26 57	27 17
5 44	26 20	24 6	23 40	25 38	28 0	28 15
5 48	27 15	25 3	24 42	26 44	29 3	29 13
5 52	28 10	26 0	25 44	27 49	0 ♒ 6	0 ♓ 11
5 56	29 5	26 57	26 47	28 55	1 8	1 7

S.T.	1	2	3	4	5	6
6 0	0 ♎ 0	27 ♎ 55	27 ♏ 49	0 ♑ 0	2 ♒ 11	2 ♓ 5
6 4	0 55	28 53	28 52	1 5	3 13	3 3
6 8	1 50	29 49	29 54	2 11	4 16	4 0
6 12	2 45	0 ♏ 47	0 ♐ 57	3 16	5 18	4 57
6 16	3 40	1 45	2 0	4 22	6 20	5 54
6 20	4 35	2 43	3 3	5 27	7 21	6 50
6 24	5 31	3 41	4 7	6 32	8 23	7 47
6 28	6 26	4 39	5 10	7 37	9 24	8 43
6 32	7 21	5 38	6 14	8 43	10 25	9 40
6 36	8 16	6 36	7 18	9 48	11 26	10 36
6 40	9 11	7 35	8 22	10 53	12 27	11 33
6 44	10 7	8 34	9 26	11 58	13 27	12 28
6 48	11 2	9 33	10 30	13 3	14 28	13 24
6 52	11 57	10 33	11 34	14 7	15 28	14 20
6 56	12 53	11 32	12 39	15 12	16 28	15 16
7 0	13 49	12 32	13 43	16 17	17 28	16 11
7 4	14 44	13 32	14 48	17 21	18 28	17 7
7 8	15 40	14 32	15 52	18 26	19 27	18 3
7 12	16 36	15 32	16 57	19 30	20 27	18 58
7 16	17 32	16 33	18 2	20 34	21 26	19 53
7 20	18 27	17 33	19 7	21 38	22 25	20 49
7 24	19 24	18 34	20 12	22 42	23 24	21 43
7 28	20 20	19 35	21 17	23 46	24 22	22 39
7 32	21 17	20 36	22 23	24 50	25 21	23 34
7 36	22 13	21 37	23 28	25 53	26 19	24 29
7 40	23 10	22 39	24 33	26 57	27 17	25 25
7 44	24 6	23 40	25 38	28 0	28 15	26 20
7 48	25 3	24 42	26 44	29 3	29 13	27 15
7 52	26 0	25 44	27 49	0 ♒ 6	0 ♓ 11	28 10
7 56	26 57	26 47	28 55	1 8	1 7	29 5
8 0	27 ♎ 55	27 ♏ 49	0 ♑ 0	2 ♒ 11	2 ♓ 5	0 ♈ 0
8 4	28 53	28 52	1 5	3 13	3 3	0 55
8 8	29 49	29 54	2 11	4 16	4 0	1 50
8 12	0 ♏ 47	0 ♐ 57	3 16	5 18	4 57	2 45
8 16	1 45	2 0	4 22	6 20	5 54	3 40
8 20	2 43	3 3	5 27	7 21	6 50	4 35
8 24	3 41	4 7	6 32	8 23	7 47	5 31
8 28	4 39	5 10	7 37	9 24	8 43	6 26
8 32	5 38	6 14	8 43	10 25	9 40	7 21
8 36	6 36	7 18	9 48	11 26	10 36	8 16
8 40	7 35	8 22	10 53	12 27	11 33	9 11
8 44	8 34	9 26	11 58	13 27	12 28	10 7
8 48	9 33	10 30	13 3	14 28	13 24	11 2
8 52	10 33	11 34	14 7	15 28	14 20	11 57
8 56	11 32	12 39	15 12	16 28	15 16	12 53
9 0	12 32	13 43	16 17	17 28	16 11	13 49
9 4	13 32	14 48	17 21	18 28	17 7	14 44
9 8	14 32	15 52	18 26	19 27	18 3	15 40
9 12	15 32	16 57	19 30	20 27	18 58	16 36
9 16	16 33	18 2	20 34	21 26	19 53	17 32
9 20	17 33	19 7	21 38	22 25	20 49	18 27
9 24	18 34	20 12	22 42	23 24	21 43	19 24
9 28	19 35	21 17	23 46	24 22	22 39	20 20
9 32	20 36	22 23	24 50	25 21	23 34	21 17
9 36	21 37	23 28	25 53	26 19	24 29	22 13
9 40	22 39	24 33	26 57	27 17	25 25	23 10
9 44	23 40	25 38	28 0	28 15	26 20	24 6
9 48	24 42	26 44	29 3	29 13	27 15	25 3
9 52	25 44	27 49	0 ♒ 6	0 ♓ 11	28 10	26 0
9 56	26 47	28 55	1 8	1 7	29 5	26 57
10 0	27 ♏ 49	0 ♑ 0	2 ♒ 11	2 ♓ 5	0 ♈ 0	27 ♈ 55
10 4	28 52	1 5	3 13	3 3	0 55	28 53
10 8	29 54	2 11	4 16	4 0	1 50	29 49
10 12	0 ♐ 57	3 16	5 18	4 57	2 45	0 ♉ 47
10 16	2 0	4 22	6 20	5 54	3 40	1 45
10 20	3 3	5 27	7 21	6 50	4 35	2 43
10 24	4 7	6 32	8 23	7 47	5 31	3 41
10 28	5 10	7 37	9 24	8 43	6 26	4 39
10 32	6 14	8 43	10 25	9 40	7 21	5 38
10 36	7 18	9 48	11 26	10 36	8 16	6 36
10 40	8 22	10 53	12 27	11 33	9 11	7 35
10 44	9 26	11 58	13 27	12 28	10 7	8 34
10 48	10 30	13 3	14 28	13 24	11 2	9 33
10 52	11 34	14 7	15 28	14 20	11 57	10 33
10 56	12 39	15 12	16 28	15 16	12 53	11 32
11 0	13 43	16 17	17 28	16 11	13 49	12 32
11 4	14 48	17 21	18 28	17 7	14 44	13 32
11 8	15 52	18 26	19 27	18 3	15 40	14 32
11 12	16 57	19 30	20 27	18 58	16 36	15 32
11 16	18 2	20 34	21 26	19 53	17 32	16 33
11 20	19 7	21 38	22 25	20 49	18 27	17 33
11 24	20 12	22 42	23 24	21 43	19 24	18 34
11 28	21 17	23 46	24 22	22 39	20 20	19 35
11 32	22 23	24 50	25 21	23 34	21 17	20 36
11 36	23 28	25 53	26 19	24 29	22 13	21 37
11 40	24 33	26 57	27 17	25 25	23 10	22 39
11 44	25 38	28 0	28 15	26 20	24 6	23 40
11 48	26 44	29 3	29 13	27 15	25 3	24 42
11 52	27 49	0 ♒ 6	0 ♓ 11	28 10	26 0	25 44
11 56	28 55	1 8	1 7	29 5	26 57	26 47

We see that the Morin Point for Prince Harry is 17°13' Aquarius, etc. Now we are ready to insert the positions of the planets by referring back to the September ephemeris of 1984. We need to make small corrections to the positions given for midnight at the beginning of the 15th by using those at the beginning of the 16th and correcting for the differences between them by multiplying these by 15 and a third and dividing by 24 (hrs). The results added on to the positions at midnight on the 15th give us the required positions at the time of the Prince's

74

birth. The Moon's position is the important one to carry out the correction for because it moves the most quickly through the Zodiac. We can fill in the aspects (with their respective orbs: + or - 8^0 for conjunction, opposition, trine and square; 4^0 for sextile and 2^0 for minor aspects) that each planet receives from the others, almost by inspection, and we can also add the Dragon's Head as well as the Part of Fortune (Morin Point – Sun + Moon). For completeness, planetary declinations[29], as well as the various parallels of declination that result, have been inserted, although these are rarely used for interpretation purposes.

The calculations at the top left (day for a year progressions) could be used to indicate future trends for Prince Harry for the year 2000. Below that we see that his ruling planet is Uranus (rules Aquarius) in the 11th House and there are various other indicators below that, useful for interpretation, as are also found in the notes column immediately to the right of the aspects table.

To generate a good indication for the whole interpretation a substantial number of these indicators together with their interpretations* is now presented:

Prince Harry: Character, Needs and Goals.

Indicator	Interpretation
Bucket Shaping, Moon Handle, behind.	Ability to influence people. Adaptable. More interested in means than ends. At best, a leader/teacher. At worst, an agitator or malcontent. Intense and conservative. See Ref. 29c.
Planets South See Ref. 17.	Preoccupation with externals, objective viewpoint, and a type of extraversion.
Lack of Air See Ref. 16.	Lack of idealism, respect for the abstract, spiritual and the theoretical.

Mainly succedent.	Likely to preserve status. See Ref. 16.
6 -ve; 4 +ve.	Somewhat self-repressive and receptive. See Ref. 16.
Not well integrated.	May have difficulty in establishing effective links between various parts of his being and awareness. See Ref. 16.
No retrograde planets.	Should have no difficulty keeping his attentions firmly fixed in the present. See Ref. 3.
Saturn unaspected.	Tendency either to all caution/control or none. Erratic self-control, not always serious and responsible, lacks self-discipline. See Ref. 3.
Sun is mutable, Moon is fixed, Morin Point is fixed. 2 fixed, 1 mutable. See Ref. 29a.	Strength of will, power and resolution, but of a quiet kind. "Looks before and after" readily and so displays foresight and method. Plans undertakings small and great, for weeks, months and years ahead and then carries them out patiently, persistently and unchanged. What may be lacking in brilliancy of thought/action is made up for in patience, forethought, steadfastness and strength of will. Feelings and affections strong and lasting, but slow to be roused, and not readily expressed. Lacks spontaneity, but much staying power. Can be discouraged by circumstances, but recovers and works on as before. Can miss opportunities through hesitation, caution or lack of receptivity.
Mutable and fixed Earth synthesis. See Ref. 29a.	Nervous temperament, methodical, slow and reserved method of expression. Life too formal or monotonous. Commonplace business habits such as shop-keeping, bookbinding or upholstery. Lack of power for full self-expression. Critical attitude of mind stimulates him and helps to break down the rigid and reserved tendency of the body. May lack opportunity owing to a timorous, fearful nature -- dreading results and hesitating to take risks. Materialistic mind, reserve, obstinacy and

75

	persistence. Over confident, uncompromising, lacks self-esteem, rigid and inflexible. Methodical and over cautious. Favours farmers, land-owners, labourers – limitations of fate are marked.
Sun-Moon Polarity. Sun in Virgo - Moon in Taurus. See Ref. 29a.	Intuitive mind with ability to take care of both body and brain. Keenly sensitive to the unspoken thoughts and intentions of others, has insight into the future of business transactions, being rarely surprised at the result. Too active for his own good, his energy exceeding his strength, but his keen perception of hygiene tends greatly to counter this. Good practical business ability and commonsense. Born into a prosperous environment.
Sun–Moon polarity. Sun in 8th– Moon in 4th. See Ref. 29b.	Your early conditioning helped your personality development and allowed you to succeed and be fulfilled. Because of your harmonious temperament you can cope with frustration while establishing your position in the world You know how to adapt to changes with favourable results, but you may not be as aggressive as you should be. You often let matters wither away because you fail to act decisively, perhaps because you feel that a better situation may develop. Unless you work out a plan for gaining the goals you want and are willing to persist in that effort, they won't materialise. You may be overly influenced by your parents expectations, but you can make your own decisions. Property, financial investment, physical therapy or medicine are some of your possible outlets for creativity. With your understanding and compassion for people you will probably attract a partner who needs your support, both personally and professionally, to achieve a

	harmonious relationship. But you may feel that your mate doesn't give you the support you need in your goals.
Aquarius rising, 2nd decanate. See Ref. 29a	Mind is active and inclines towards intellectual attainment but does not have sufficient continuity, or concentration, to qualify for any literary or educational work, favouring mechanical and physical activities in the business world, more than mental pursuits. The fate is often affected by relatives and companions.
Uranus in Sagittarius in 11th House – ruler. See Ref. 16.	Desire for enlightenment, spiritual aspirations, prophetic insight, experimental tendencies, courage to rebel, awakened. Conventionality is destroyed with a spirited will. Tends to be excitable and utopian. Original aspirations, advanced ideals; remarkable friendships, sudden attachments, estrangements, unconventional, independent, inventive, occultist friends. Though conducive to strong mental action through revolutionary thought, communicativeness is too brusque and independent so that it loses good contact with others. The addiction to the unusual and unconventional is so strong and so awkwardly expressed that the person becomes eccentric, odd and tiresome.
4th from 8th. See Ref 29b.	You are conscious of your obligations to society, and you know you must serve others' needs in order to justify your claim to success fully. Helping others get started in their endeavours should be rewarding.
Part of Fortune, in Libra in 9th, sext. Mars.	You will feel pleased when you contribute to the well-being of an active group. This will also generate in you a sense of belonging. Truth discovered in the process will also give you pleasure. See Ref. 29d.

Having obtained all the required interpretations the next task is to divide it all (leaving out the indicators) into four sections, namely characteristics, relationships, career and health. Further subdivisions may be suitable. In this way we produce a "synthesised", whole interpretation ready for submitting to the client (Prince Harry).

- -

References and footnotes: (lower numbered references occur at the ends of previous chapters/appendices.)

28) The American ephemeris for the 20th Century, 1900 – 2000, at midnight, N. F. Michelson, Astro-Computing Services, San Diego, USA, 1980.

29) Raphael's Ephemeris for 1984, W. Foulsham & Co. Ltd., Slough, England.

*   Suggested sources of interpretations include references 16, 17, 22, and 23 as well as:-

a) A. Leo, "Astrology for All", Samuel Weiser, New York, U.S.A., 1978.

b) R. Pelletier, "Planets in Houses" and "Planets in Aspect", Para research Inc., Rockport, Mass., U.S.A., 1978 and 1974.

c) M. E. Jones, "The Guide to Horoscope Interpretation", The Theosophical Publishing House, Wheaton, Illinois, U.S.A., 1974.

d) M. Schulman, "Karmic Astrology. Vols. 1 – 4", The Aquarian Press, Wellingborough, U.K., 1978-9.

Astrological Computer software often comes complete with its own series of interpretations.

- -

APPENDIX 5b

The Mars Effect and the Morin Point

Astrologically, the interpretation of the planet Mars in horoscopes involves words synonymous with heat, energy and action[19]. Consequently we should expect that Mars should feature strongly in the horoscopes of sports champions. Using statistical methods the Gauquelins rigorously and objectively detected a Mars Effect[30] in the horoscopes of sports champions, which Ertel[6] has confirmed. However, they devised their own system of sector distributions in order to detect their Mars effect without statistical bias, but also without astrological logic. They noted that there was little statistical evidence for the existence of House Systems based on the traditional Ascendant, or even for Zodiacal Signs. On the other hand the simple Morinus House System, with its own logic and world-wide application, has its own specific "Ascendant" (the Morin Point) that is not too far removed from the "Oblique Ascension" used by the Gauquelins. By using the Morin Point, together with the Gauquelins' 450 Serie D New Sports Champions file[31], evidence was found for a Mars Effect. Polar Graph A shows the sector positions (1 – 36) of the planet Mars from the Morinus horoscopes of the 450 New Sports Champions:

The circumference of the circle gives the average number expected (12.5) for a uniform distribution. Areas greater than expectation (dark shading) are found clustered around the Morin Point in particular, but also near the Morinus "Midheaven" (close to the traditional one), as well as to its right. Chi-squared values for sectors 36, 1, 2 and 3 are all greater than 10. This tends to confirm that the greater area around sectors 36, 1, 2 and 3 is genuine.

Polar Graph A: The Sector Positions (1 – 36) of the Planet Mars in the Horoscopes of the 450 New Sports Champions (Gauquelins' Serie D file).

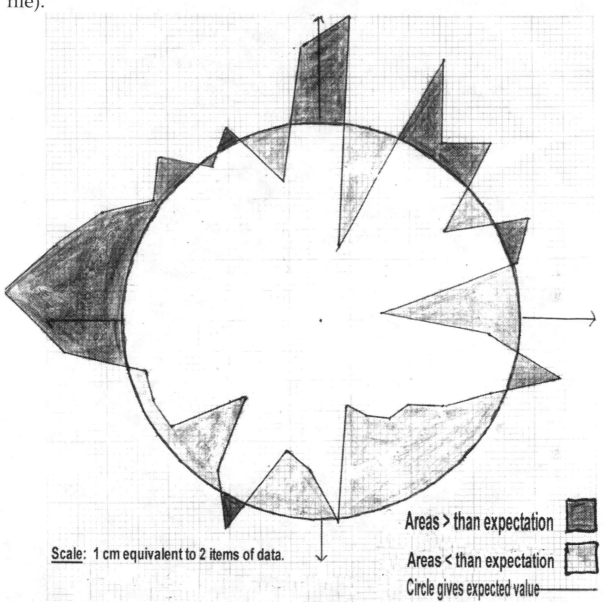

Scale: 1 cm equivalent to 2 items of data.

Areas > than expectation

Areas < than expectation

Circle gives expected value

Superficially the graphical distribution found by the Gauquelins, and that shown by Polar Graph A. look similar, but lesser areas (light shading) in A extend all the way beneath the horizontal axis rather than showing other noteworthy greater areas near to the Descendant (right axis) or to the I.C. (bottom axis). Actually there are substantially more positions above the Morin Point (253), indicating action (from

Mars) objectivity in A than below it (197) and also to the East, indicating action (from Mars) and destiny decision, compared with the West.

An attempt to replicate these results using the Gauquelins' 2087 Serie A Sports Champions file, led to Polar Graph B.

<u>Polar Graph B</u>: The Sector Positions (1 – 36) of the Planet Mars in the Horoscopes of the 2087 Sports Champions (Gauquelins' Serie A file).

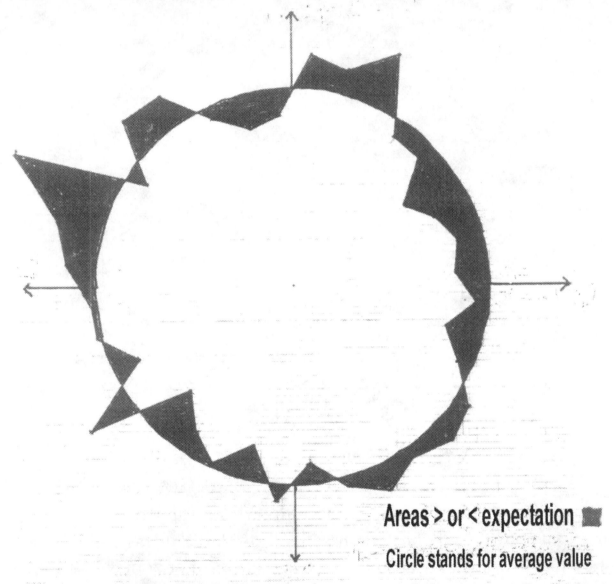

Areas > or < expectation ■

Circle stands for average value

<u>Scale</u>: 1cm equivalent to 10 items of data.

The circle itself gives the average number (58) for a uniform distribution. Areas greater than expectation are found above the

Morin Point and round beyond the Midheaven. These areas are significant (X^2 sector 3 = 17.6) statistically due to the larger sample size (p% < 0.5). Areas less than expectation are found surrounding the Morinus Descendant (OMoPo) (right axis) and round beyond the I.C. (bottom axis). The positions of the greater areas are roughly in line with those the Gauquelins found for their own sector distributions for serie A, but the overall shape and positions of the maxima and minima in Polar Graph B are significantly different from those in Polar Graph A.

The results suggest (cautiously) that a) the champions in Serie D are more eminent than those in Serie A and b) the timings of birth used for both Serie are not sufficiently reliable/accurate for conducting quantitative studies properly.

There is a relatively minor problem met by using the Morinus House System directly for statistical analysis. Equidistant sectors along the equator suffer certain, regular distortion when projected onto the ecliptic (from the poles of the ecliptic) so that, depending on the position of the Morin Point, sectors will be somewhat larger or smaller (up to 10%) than average. However, provided that the distribution of the Morin Points of the sports champions is uniform throughout the Zodiac, then differences due to sector size will largely cancel out. The Bar Chart shows the distribution of the Morin Point among the Signs of the Zodiac in the daily circle for the horoscopes of the Serie A Sports Champions:

The horizontal dashed line gives the average number expected (174) for a uniform distribution. Comparison with a similar graph for Serie D champions shows that, although the maxima and minima are essentially (only one different) replicated, the differences between them are nothing like so large (these results for the Bar Chart for Serie A are significant only at the 12% level, for Serie D, 2.5%). This result supports the assumption that the Morin Point of the Sports

Champions is practically uniform throughout the Zodiac. Presumably the larger sample size of Serie A has ironed out the differences.

<u>Bar Chart</u>: The Distribution of the Morin Point among the Signs of the Zodiac for the Horoscopes of the Gauqulins'2087 Serie A Sports Champions.

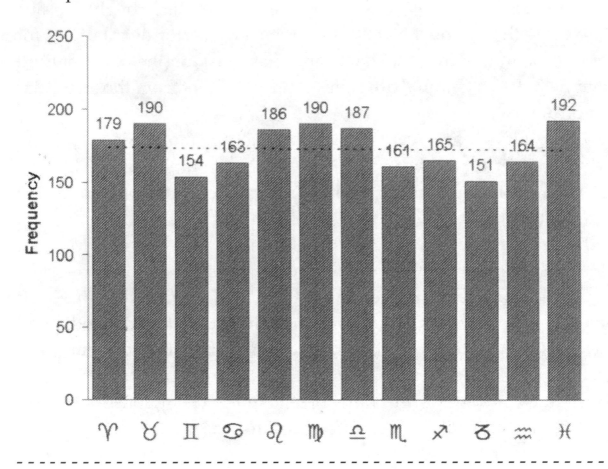

References and footnotes: (lower numbered references occur at the ends of previous chapters.)

30) Planetary Heredity, M. Gauquelin, A.C.S. Publications, San Diego, USA, and references cited there.

31) The Sports Champions files were obtained from C.U.R.A. (http://cura.free.fr), 2002. I have assumed that the Zelen test (use of control samples) would hold for this work.

APPENDIX 6

Determination of an Ideal Birth Moment.

The birth of a girl took place in Rock Ferry, Wirral, England on the 30th August, 1936 at 09:30 p.m. clock time, i.e. at 08:30 p.m. after subtracting one hour of British summertime. This gives a local sidereal (star) time at birth of 18 hrs. 53 mins, at which, according to Tables of Morinus Houses (see Appendix 5a) 12^0 15′ Aries (the Morin Point) was rising. On this date, and at this time, the Moon was situated at 14^0 35′ Aquarius. Consulting Bailey's Table of Sex Degrees (see Appendix 7) we see that the Morin Point lies in a male area, and that the Moon lies in a negative (non-sex) one. The Moon lies above the Morin Point (see the Diagram), and because the Sun is at 7^0 15′ Virgo, the Moon is therefore increasing. This means that we have an Epoch of the 1st. Order. Nine months earlier (i.e. ten cycles of the Moon earlier) the Moon was situated at 14^0 35′ Aquarius on 1st Dec. 1935. This is the Index Date. The Moon crosses 12^0 15′ Libra on the 19th Dec. 1935. This is the day of the Epoch. The actual Epoch will be either 2nd or 3rd variation irregular[1].

For the 2nd Irregular Epoch on 19/12/1935 14^0 35′ Aquarius rises in Rock Ferry at time T, where T is given by the equation:

$$5:47 + T + A - 00:12 = 14:48$$

$$\text{(S)} \qquad \text{(R)} \quad \text{(L.S.T.)}$$

S is the sidereal (star) time at midnight at the start of the day, A is the adjustment between clock time and sidereal time (9.86 secs. per hour), R is the longitude correction and L.S.T. stands for local sidereal time. From this equation we find that T + A = 9 hrs. 13 mins. Now at this time A = 1.5 mins., so that T = 9 hrs. 11.5 mins. On this date and at this time the Moon is situated at 12^0 31′ Libra. Here the Moon lies in a female sex position, in agreement with that for a girl birth. On the day of birth (30/8/1936) this position of the Moon sets (12^0 31' Aries rises) at time T, where T is given by the equation:

$$22:32 + T + A - 00:12 = 18(42):54.5$$

From this we find that $T + A = 20:34.5$. At this time $A = 3.5$ mins., so that $T = 20:31$. With the hour added back for summertime this gives a clock time of 09:31 p.m. on 30/8/1936.

For the 3rd irregular Epoch we also find a birth time for a girl, but this takes place about 25 mins. later.

To obtain an ideal birth time to the nearest second, we need to repeat the procedure, but more accurately, as follows:

At 09:31:00 p.m. on 30/8/1936 the position of the Moon is 14^0 35' 44" Aquarius. On the 19/12/1935 14^0 35' 44" Aquarius rises at Rock Ferry (lat. 53^0 22' 05" N; long. 3^0 00'12" W) at time T, where T is given by:

$$5:46:40 + T + A - 00.12:01 = 14:48:31.$$

From this we find that $T + A = 09:13:52$. For this time $A = 00:01:31$ so $T = 09:12:21$. At this time the Moon's position is 12^0 29' 34" Libra. On the day of birth (30/8/1936) the Moon's position sets (12^0 29' 34" Aries rises) at time T, where T is given by:

$$22:32:02 + T + A - 00:12:01 = 18(42):54:20$$

From this we find that $T + A = 20:34:19$; and $A = 00:03:23$ so that $T = 20:30:56$. At this time the Moon's position is 14^0 35' 41" Aquarius. Clock time requires that one hour of summertime is added onto T. For this case a second iteration, using these more accurate figures, serves only to confirm the foregoing birth time of 20:30:56. Addey[7] has argued that birth times of this sort of accuracy are required for high number (e.g. 47th) harmonic analysis of birth charts.

Notice that the Moon's position in the epoch chart is at the opposite of that of the Morin Point in the birth chart and that the Moon's position in the birth chart is at that of the Morin Point in the epoch chart. Notice also that the Moon's position at birth is a reflection of that at epoch about the vertical (M.C.-I.C.), plane mirror.

Birth Chart and Aspect Grid for Joan M.

Joan M—Birth
Female Chart
30 Aug 1936
21:30:56 BST −1:00
Rock Ferry, England
53°N22' 002°W59'
Geocentric
Tropical
Morinus
Mean Node

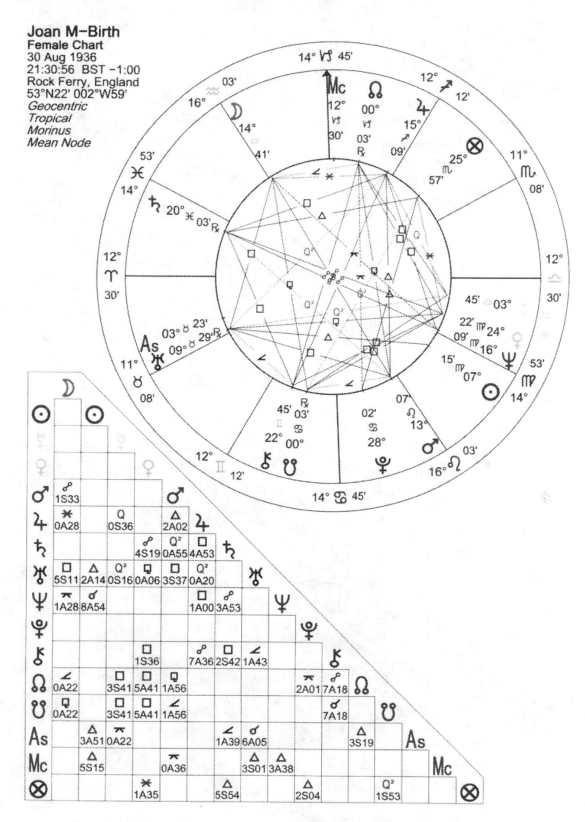

Epoch Chart and Aspect Grid for Joan M.

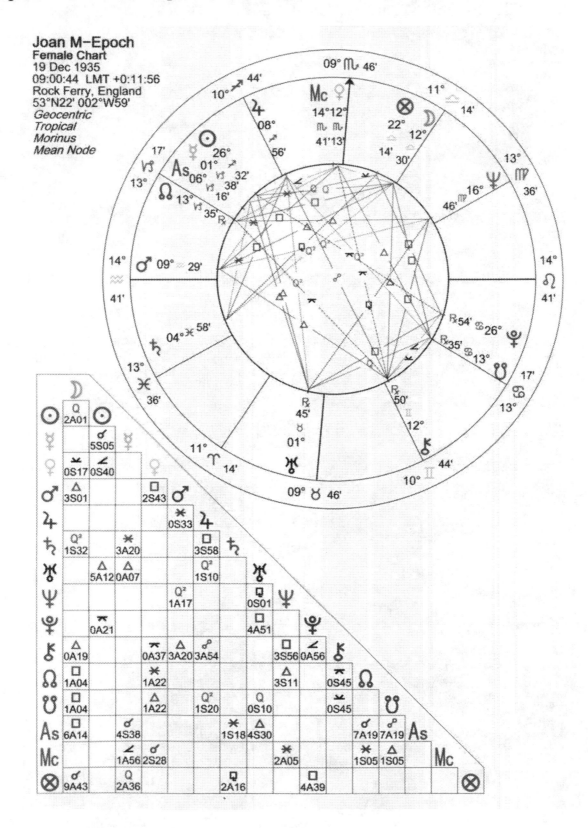

Joan M-Epoch
Female Chart
19 Dec 1935
09:00:44 LMT +0:11:56
Rock Ferry, England
53°N22' 002°W59'
Geocentric
Tropical
Morinus
Mean Node

Appendix 7

The First and Second Cases

<u>The first case</u>

Beginning at midnight at the start of the 8th February, the Moon was at 26⁰42' Cancer and at noon that day was at 3⁰25' Leo. Examination of Bailey's Table of Sex Degrees shows that before insemination, and indeed for the following two days, the Moon was in a male sex area. This means that not only does the Morin Point have to be in a female sex position, but that the Moon has to be in a female quadrant also. The Morin Point* at midnight at the beginning of the 8th February, 1990 on the Wirral, was 12⁰30' Scorpio, a male position. The Morin Point lies within a female area between 17⁰09' and 25⁰43' Scorpio, i.e. between 00:11 a.m. and 00:52 a.m. on that day. At the exact sex point (21⁰26' Scorpio rising) the position of the Moon is 27⁰ Cancer. This means that on a prospective date-of-birth the Moon will be roughly 21⁰30' Scorpio (or Taurus), and that 27⁰ Cancer (or Capricorn) will be rising. Following inspection of the Ephemeris[28], possible dates for consideration are therefore the 20th October, the 3rd November and the 17th November, 1990.

For the 20th October we find that, when 27⁰ Cancer rises, the Moon is below the Earth and increasing, so that the order of the Epoch is 3. As the Index Date is the 21st January, this particular birth chart cannot be valid because its Epoch must be before the 21st January, and so cannot occur on the 8th February. When 27⁰ Capricorn rises on the 20th October the Moon is above the Earth and increasing so that the order of the Epoch is 1. Both the Moon and the Morin Point lie in a female sex area. Since the Index Date is the 21st January we require an irregular 2nd variation Epoch[1] in order to obtain an Epoch on the 8th February. Here the Moon must lie in a female quadrant, which it does. Hence 27^0 Capricorn rising on the 20th October, 1990 provides us with one, just possible, birth date and time[17], for a girl baby.

Consider now the 3rd November, 1990, in the afternoon, when the Moon will be at $21^026'$ Taurus. 27^0 Cancer rises on this date at 11:01 p.m. At this time the Moon lies at $26^009'$ Taurus, which is just outside the allowed area. Additionally the Moon is above the Earth and decreasing so that the Order of the Epoch is 2. The Index Date is the 3rd February, 1990, which cannot produce a 2nd Order Epoch on the later date of the 8th February.

27^0 Capricorn rises on the 3rd November, 1990 at 11:02 a.m. At this time the Moon lies at $18^036'$ Taurus. The Moon is below the Earth and decreasing so that the Order of the Epoch is 4. As the Index date is the 3rd February, we can have an Epoch on the 8th February following. When $18^034'$ Scorpio rises on the 8th February, 1990 we find that the Moon's position is $26^050'$ Cancer. The Morin Point is female, the Moon is male and the Moon lies in a female quadrant. Here we have found a valid Epoch for a girl baby to be born at 11:02 a.m. on the 3rd November, 1990 on the Wirral.

Now consider possibilities for birth on or around the 17th November, 1990 on which day the Moon is at $21^026'$ Scorpio early in the morning. 27^0 Cancer rises on the 16th November at 10:09 p.m. at which time the Moon is at $19^018'$ Scorpio. The Moon is below the Earth and decreasing so that the Order of the Epoch is 4. The Index Date is the 17th February, 1990. This means that the Epoch must come after the 17th February, so that this potential birth-time is invalid. 27^0 Capricorn rises on the 17th November, 1990 at 10:07 a.m. at which time the Moon is at $25^017'$ Scorpio. The Moon is above the Earth and increasing so that the Order of the Epoch is 1. Since the Index Date is the 17th February, 1990 this means again, that the Epoch must come after 17th February, so that this potential birth-time, too, is invalid. Any other position of the Moon around this time falls outside the allowed female range ($17^009'$ - $25^043'$ Scorpio).

On the 8th February, 1990 (the day of Epoch) the Morin Point re-enters a female area at 21°26 Capricorn i.e. at 4:20 a.m. At this time the position of the Moon is 29°07' Cancer. Consulting the Ephemeris[28] shows us that possible dates of birth for a girl are the 26th October (late 25th?), the 8th November and the 22nd November. Proceeding in exactly the same way as just described, we find that there cannot be a suitable birth-date late on the 25th October. This is because the Index Date occurs on the 25th January. Because the Order of the Epoch is 3 it would have to occur before that date, and so cannot occur on the 8th February. However, an interesting situation arises when 29°07' Capricorn ascends on the 26th October, at 11:34 a.m. At this time the position of the Moon is 28°41' Capricorn, very close to the Morin Point. As the Moon is above the Earth (just) and increasing, the Order of the Epoch is 1. The Index Date is the 26th January so that this must be a case of Moon to the Morin Point and round to the opposite of the Morin Point to derive an Epoch on the 8th February[1]. We need then to follow the rules for an irregular Epoch, 2nd variation[1]. Thus 28°41' Capricorn rises on the 8th February at 4:45 a.m. At this time the Moon is at 29°23' Cancer. The Moon is male, the Morin Point is female and the Moon is just inside a female quadrant. So here we have a potential time for a future girl birth.

Considering possible girl births on the 8th November, 1990 we find either that the Epoch must be before the Index Date of the 7th February, or that it must be 28 days before the Index Date. Similarly for possible girl births on the 22nd November, 1990 we find that valid Epochs need to occur after the Index Date of the 22nd February.

Immediately after 4:45 a.m. on the 8th February we find that the Moon remains in a male quadrant for nearly 7 hours, effectively prohibiting the possibility for Epochs leading to girl births during this time. The next time that both the Morin Point and the quadrant containing the Moon are female occurs when 4°17' Taurus rises on the 8th February at 11:25 a.m. on the Wirral. This is now more than twelve hours after insemination and likely to be beyond the time required between insemination and fertilisation. Hence we are left with only the three possible Epochs as given, leading to the birth of a girl baby some nine months later.

The second case

Proceeding as directed by the stepwise summary we find from Bailey's Table of Sex Degrees that the area occupied by the Moon at Epoch, and for all that day, was masculine. Once again, not only does the Morin Point have to be in a female sex area at Epoch, but that the Moon has to be in a female quadrant. The Morin Point enters a female sex area when $0^0 00'$ Gemini rises on the 28th April, 1986. The Moon will enter a male quadrant when the Morin Point becomes 2^0 Cancer. There is thus a period of two hours starting at 8:55 a.m. during which Epochs for girl births become possible. The next time that both the area occupied by the Morin Point, and the quadrant containing the Moon, are female, occurs when $8^0 34'$ Libra rises at 5:25 p.m. that day. This is almost nine hours after insemination took place. Hence we can sensibly confine our attention to the two-hour period between 8:55 and 10:55 a.m. on the day of Epoch. During this time the Moon, on average, is at $1^0 15'$ Capricorn. This means that on a prospective date-of-birth the Moon will be between 0^0 Gemini and 2^0 Cancer (or between 0^0 Sagittarius and 2^0 Capricorn) and $1^0 15'$ Capricorn (or $1^0 15'$ Cancer) will be rising. Following inspection of the Ephemeris[28], possible dates of birth for consideration are 10th-13th January (before the Caesarian birth!)+, 25th – 27th January, and the 7th– 9th February, 1987. Surprisingly, no suitable birth-time occurs between the 25th– 27th January. This is either because the Epoch is 2nd Order and so must occur before 28th April, 1986, or because whenever the Epoch is 4th Order, the Moon falls in a negative area and so cannot generate an Epoch for a girl birth. The three potential birth times having $1^0 15'$ Cancer rising on the 7th, 8th and 9th February, lead to 1st Order Epochs that must fall after the Index Date of the 10th May, 1986. However, the three potential birth-times having $1^0 15'$ Capricorn rising on these dates give 3rd Order Epochs leading to three valid Epochs for girl births on the required date of the 28th April, 1986. Note that, once again, we are left only to deal with three, potential, speculative birth-times.

BAILEY'S TABLE OF SEX DEGREES

Sex.	Limits of		Exact Sex Point.	Limits of	
	Moon's Orb.	Morin Pt's Orb.		Morin Pt's Orb.	Moon's Orb.
F	—	—	♈ 0.0	♈ 4.17	♈ 6.26
M	♈ 6.26	♈ 8.34	♈ 12.51	♈ 17.9	♈ 19.17
M	♈ 19.17	♈ 21.26	♈ 25.43	♉ 0.0	♉ 2.9
F	♉ 2.9	♉ 4.17	♉ 8.34	♉ 12.51	♉ 15.0
M	♉ 15.0	♉ 17.9	♉ 21.26	♉ 25.43	♉ 27.51
F	♉ 27.51	♊ 0.0	♊ 4.17	♊ 8.34	♊ 10.43
F	♊ 10.43	♊ 12.51	♊ 17.9	♊ 21.26	♊ 23.34
F	♊ 23.34	♊ 25.43	♋ 0.0	♋ 4.17	♋ 6.26
F	♋ 6.26	♋ 8.34	♋ 12.51	♋ 17.9	♋ 19.17
M	♋ 19.17	♋ 21.26	♋ 25.43	♌ 0.0	♌ 2.9
M	♌ 2.9	♌ 4.17	♌ 8.34	♌ 12.51	♌ 15.0
M	♌ 15.0	♌ 17.9	♌ 21.26	♌ 25.43	♌ 27.51
M	♌ 27.51	♍ 0.0	♍ 4.17	♍ 8.34	♍ 10.43
F	♍ 10.43	♍ 12.51	♍ 17.9	♍ 21.26	♍ 23.34
M	♍ 23.34	♍ 25.43	♎ 0.0	♎ 4.17	♎ 6.26
F	♎ 6.26	♎ 8.34	♎ 12.51	♎ 17.9	♎ 19.17
F	♎ 19.17	♎ 21.26	♎ 25.43	♏ 0.0	♏ 2.9
M	♏ 2.9	♏ 4.17	♏ 8.34	♏ 12.51	♏ 15.0
F	♏ 15.0	♏ 17.9	♏ 21.26	♏ 25.43	♏ 27.51
M	♏ 27.51	♐ 0.0	♐ 4.17	♐ 8.34	♐ 10.43
M	♐ 10.43	♐ 12.51	♐ 17.9	♐ 21.26	♐ 23.34
M	♐ 23.34	♐ 25.43	♑ 0.0	♑ 4.17	♑ 6.26
M	♑ 6.26	♑ 8.34	♑ 12.51	♑ 17.9	♑ 19.17
F	♑ 19.17	♑ 21.26	♑ 25.43	♒ 0.0	♒ 2.9
F	♒ 2.9	♒ 4.17	♒ 8.34	♒ 12.51	♒ 15.0
F	♒ 15.0	♒ 17.9	♒ 21.26	♒ 25.43	♒ 27.51
F	♒ 27.51	♓ 0.0	♓ 4.17	♓ 8.34	♓ 10.43
M	♓ 10.43	♓ 12.51	♓ 17.9	♓ 21.26	♓ 23.34
F	♓ 23.34	♓ 25.43	♓ 30.0	—	—

EXPLANATION.—Column 4 of the table shows the exact sex point. Cols 3 and 5 show where the influence of the Morin Pt commences and finishes respectively. Cols. 2 and 6 show where the moon's influence commences and finishes respectively. Col. 1 gives the sex of the area between the longitudes in Cols. 2 and 6.

EXAMPLE.— Morin Pt at epoch, Cancer 18° 5'. Moon, Scorpio 15° 24'. Looking down the columns headed "Morin Pt.'s Orb," it will be seen that Cancer 18° 5' is outside the limits given. It is, therefore, negative. Looking down the columns headed " Moon's Orb," it will be seen that Scorpio 15° commences a male area, and the moon will therefore be in an area of that sex. The same rule applies to all cases.

BAILEY'S SEX QUODRANTS

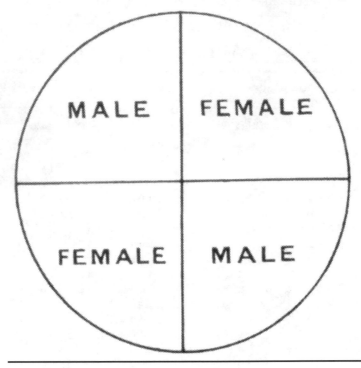

- -

APPENDIX 8

A Short Review of Planetary Heredity

Addey[7] investigated planetary heredity in the Zodiacal circle, using midpoints, quintiles and their sub-harmonics, with the M.C. merely as a reference point. He maintained that the laws of genetic astrology needed discovering so that we can explain order down the generations. On the distaff side, the stability of Moon positions (again in the Zodiac), among the womenfolk, was examined. For example a son marries a girl with the same Moon position, or its opposite, as that of his mother, and so on. Addey[7] also investigated small harmonics in the Zodiac such as the 46th (7°50') that appeared to be characteristic for a particular family. Harvey[32] suggested that children could often seem more like their grandparents than parents, i.e. inherited characteristics had skipped a generation.

Close aspects to the Sun, Moon and ruler (in the aspect circle, in the Zodiac) are important. Jayne and Schaeffer[33] analysed Mars aspects for 549 French athletes (ex the Gauquelins) and found them to be 12% more numerous than expected. The most significant planet was Jupiter, and then Venus. The more significant aspects were the square, the sextile and then the trine.

Originally the Church of Light[34], Choisnard[35] and Le Clerq[36] separately looked for heredity factors in the daily circle providing results similar to those of the Gauquelins, but their techniques were questionable. The Gauquelins[30] found that parents and children tended to have the same angular planets. They found that geomagnetic activity enhances the effect, but found no evidence for heredity for Zodiacal signs, for Houses or for aspects to the Sun, Moon or M.C. The reality of planetary heredity was demonstrated statistically. Hence they provided rigorous and objective evidence about the fundamentals of astrology upon which everything depends. Later on their results could not be replicated, particularly in the USA.

Others[3] claim that conjunctions and oppositions to the M.C. and to the rising point (or to points 10^0 past these points) (i.e. in the daily circle) are important.   However, the Gauquelins could not confirm these results using data from their own 2088 Serie A file.

The question to ask now is: "Can we pick out planetary heredity principles by using the Morin Point in the daily circle that may have been overlooked previously because we used the traditional (or Placidus) Ascendant?"   Examination of the contents of Chapter 8 allows us to conclude, provisionally, that the answer is "Yes".

- -

References and footnotes: (lower numbered references occur at the ends of previous chapters/appendices.)

32) C. Harvey, Astrology and Genetics, in Correlation, vol. 3, Spring 1969.

33) C. A. Jayne and R. S. Schaeffer in Aquarian Agent, USA, vol. 1, no. 9, p 13, 1970.

34) D. G. Doane Ed., Church of Light, Astrology: 30 years research, Professional Astrologers, Los Angeles, USA, 1956.

35) P. Choisnard, Preuves et Bases de l'Astrologie Scientific, Editions Chaconac, Paris, 1921.

36) G. Le Clerq in CAO Times, vol. 2 no. 3, p 7, 1976.

- -

APPENDIX 11
Chris Stubbs

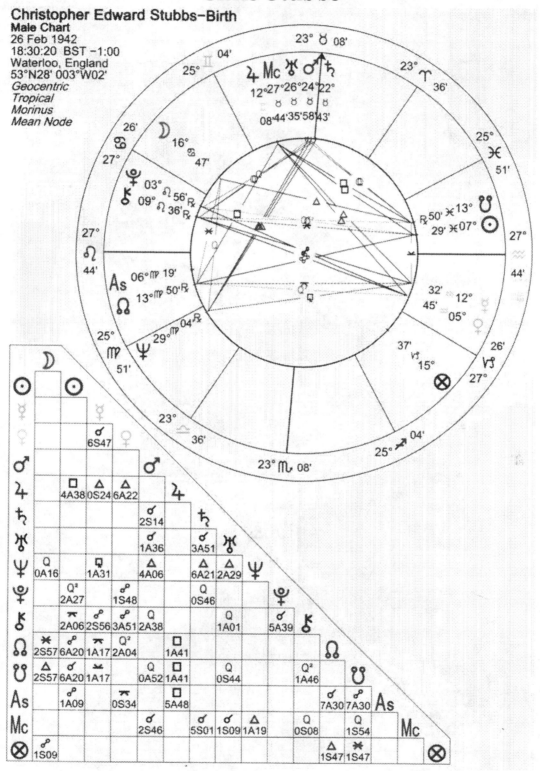

Christopher Edward Stubbs–Birth
Male Chart
26 Feb 1942
18:30:20 BST –1:00
Waterloo, England
53°N28' 003°W02'
Geocentric
Tropical
Morinus
Mean Node

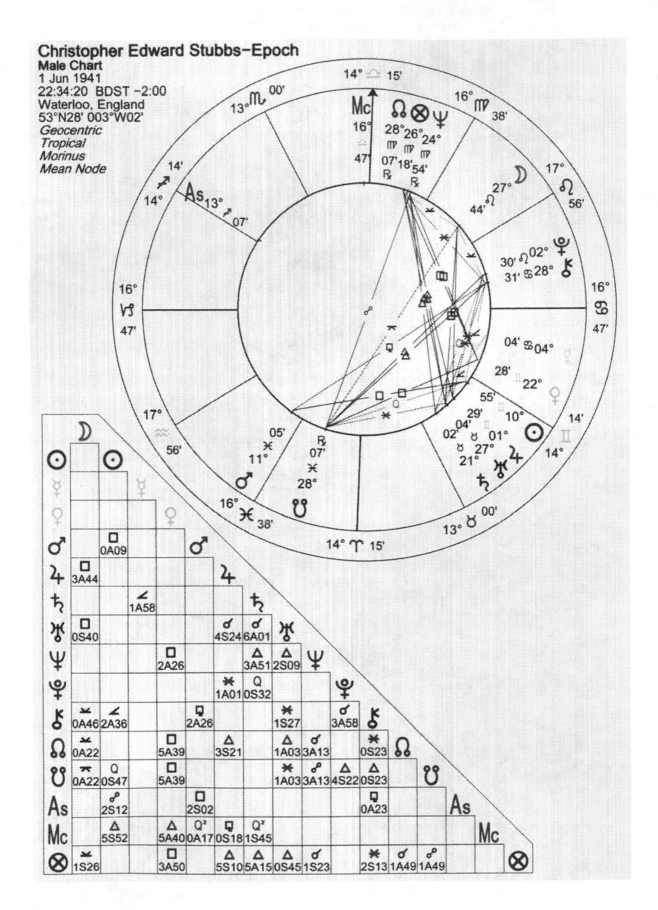

Christopher Edward Stubbs-Epoch
Male Chart
1 Jun 1941
22:34:20 BDST −2:00
Waterloo, England
53°N28' 003°W02'
Geocentric
Tropical
Morinus
Mean Node

Educated at Merchant Taylors' School for boys, Crosby, Liverpool and in Organic Chemistry at Liverpool, Chicago and Nottingham Universities. Employed by Unilever Research as a scientist both at Port Sunlight, Wirral, England and at Vlaardingen, near Rotterdam, Holland. DMS Astrol., 1984.

